Flowering and fruiting are key processes in the biology of higher plants, ensuring the transfer of genetic material from one generation to the next. In addition, as almost all of the world's agricultural and horticultural industries depend on the production of flowers, fruits and seeds, the study of the reproductive biology of cultivated plants is of fundamental importance to humankind. Surprisingly, therefore, this topic has received relatively little attention from environmental physiologists compared with studies on the growth and development of vegetative structures. This book, based on a meeting held by the Environmental Physiology Group of the Society of Experimental Biology, sets out to correct this deficiency. The topic is given a broad and comprehensive treatment, with chapters covering the onset of flowering through to the development and growth of fruits and seeds, and finally to ecological and evolutionary aspects of fruiting. This volume will therefore serve as a useful introduction to the various aspects of flowering and fruiting and will also provide a thorough general overview of the subject for students and researchers alike.

SOCIETY FOR EXPERIMENTAL BIOLOGY
SEMINAR SERIES: 47

FRUIT AND SEED PRODUCTION: ASPECTS OF
DEVELOPMENT, ENVIRONMENTAL PHYSIOLOGY
AND ECOLOGY

SOCIETY FOR EXPERIMENTAL BIOLOGY SEMINAR SERIES

A series of multi-author volumes developed from seminars held by the Society for Experimental Biology. Each volume serves not only as an introductory review of a specific topic, but also introduces the reader to experimental evidence to support the theories and principles discussed, and points the way to new research.

FRUIT AND SEED PRODUCTION
Aspects of development, environmental physiology and ecology

Edited by

C. Marshall
School of Biological Sciences, University of Wales, Bangor
J. Grace
Institute of Ecology and Resource Management, University of Edinburgh

CAMBRIDGE
UNIVERSITY PRESS

CAMBRIDGE UNIVERSITY PRESS
Cambridge, New York, Melbourne, Madrid, Cape Town, Singapore, São Paulo

Cambridge University Press
The Edinburgh Building, Cambridge CB2 8RU, UK

Published in the United States of America by Cambridge University Press, New York

www.cambridge.org
Information on this title: www.cambridge.org/9780521373500

First published 1992
This digitally printed version 2008

A catalogue record for this publication is available from the British Library

ISBN 978-0-521-37350-0 hardback
ISBN 978-0-521-05045-6 paperback

Contents

Contributors

BOWLES, D.J.
Department of Biochemistry and Molecular Biology, University of Leeds, Leeds LS2 9JT, UK.

DUFFUS, C.M.
The Scottish Agricultural College, Edinburgh, West Mains Road, Edinburgh EH9 3JG, UK.

GOLDWIN, G.K.
Department of Horticulture, Wye College, Wye, Ashford, Kent TN25 5AH.
Present address: MAFF, Nobel House, 17 Smith Square, London SW1P 3JR, UK.

HO, L.C.
Institute of Horticultural Research, Worthing Road, Littlehampton, West Sussex BN17 6LP, UK.

KAY, Q.O.N.
School of Biological Sciences, University College of Swansea, Singleton Park, Swansea SA2 8PP, UK.

LYNDON, R.F.
Division of Biological Sciences, University of Edinburgh, King's Buildings, Mayfield Road, Edinburgh EH9 3JH, UK.

MARSHALL, C.
School of Biological Sciences, University of Wales, Bangor, Gwynedd LL57 2UW, UK.

OWENS, S. J.
Jodrell Laboratory, Royal Botanic Gardens, Kew, Richmond, Surrey TW9 3AB, UK.

PIGOTT, C.D.
University Botanic Garden, Cory Lodge, Cambridge CB2 1JF, UK.

STEPHENSON, A.G.
Department of Biology, College of Science, Pennsylvania State University, University Park, Pennsylvania 16802, USA.

WATSON, M.A.
Department of Biology, Indiana University, Jordan Hall 138, Bloomington, Indiana 47405, USA.
WOOLHOUSE, H.W.
AFRC Institute of Plant Science Research, John Innes Institute, Colney Lane, Norwich NR4 7UH, UK.
Present address: Waite Agricultural Research Institute, University of Adelaide, Glen Osmond, South Australia 5064.

Preface

Flowering and fruiting are the key processes in the biology of higher plants that ensure the transfer of genetic material from one generation to the next. Furthermore, almost the whole of the world's agricultural and horticultural industries depend upon the production of flowers, fruits and seeds, and so the reproductive biology of cultivated plants is of fundamental importance to humankind. However, it is surprising that compared with studies on the growth and development of vegetative structures, reproductive biology seems to have received somewhat less attention from environmental physiologists.

Previous meetings of the Environmental Physiology Group of the Society for Experimental Biology have considered various aspects of vegetative growth in some detail and these have resulted in SEB Seminar Series publications on The Control of Leaf Growth (Cardiff Meeting, 1984), Root Development and Function (Bangor Meeting, 1985) and on Plant Canopies (Nottingham Meeting, 1986). It thus seemed timely to devote a meeting to various aspects of reproductive growth and development and so this became the subject of the 1990 meeting of the Environmental Physiology Group at the Society's annual conference at University of Warwick. It was agreed from the outset that this topic should be approached on a broad front, from the onset of flowering to the development and growth of fruits and seeds, and finally to ecological and evolutionary aspects of fruiting. Thus future meetings of the Group can focus more narrowly on topics such as seed growth and development.

We hope this Seminar Series publication will be of interest to a wide range of pure and applied plant biologists with interests in the reproductive biology of higher plants. By its broad approach it should serve as a useful introduction to the various aspects of flowering and fruiting for final-year undergraduates and research students, and also provide a volume of interest for established plant scientists who wish to obtain a general overview of the subject.

The Warwick meeting was supported by the British Ecological Society

and ICI Agrochemicals; their financial contribution is gratefully acknowledged. We thank Dr L.D. Incoll, formerly Convenor of the Environmental Physiology Group, for his encouragement in initiating this meeting, and
Dr W.J. Davies, the current Convenor, for his support in the final stages of
its organization. Lastly, we thank all our contributors for their stimulating
presentations at the meeting and for their efforts in submitting manuscripts
with a minimum of delay.

C. Marshall (Bangor)
J. Grace (Edinburgh)

H.W. WOOLHOUSE

Plant reproductive biology: an overview

Biological principles of fruit and seed production

The life cycles of flowering plants can be generalized, with a few exceptions in which sexual reproductive capacity has apparently been lost to clonal vegetative reproduction or distorted by such devices as apomixis (Fig. 1). Within this simple framework, however, the variation is enormous: the duration of the life cycle may range from days to centuries, and the proportion of the biomass invested in reproduction varies with species, genotype, age of the plant and environmental conditions (Baker, 1972). For a given reproductive investment, species may trade off large numbers of tiny seeds from a single flower, as in many orchids, as against a few bulky seeds with large reserves from a large number of flowers, as in the pome fruits. There is abundant evidence that the evolution of these details of plant reproductive biology has been influenced by a wide variety of physical factors such as availability of nutrients, light and water (Mooney, 1972) and biological factors such as the nature, availability and energetics of pollinators (Heinrich, 1975), the agents of fruit and seed dispersal (Gautier-Hion et al., 1985) and the activity of pathogens and predators (Janzen, 1977). One may trace in the literature of plant reproductive biology a process of gradual description and elucidation of factors which may influence reproductive behaviour (Lloyd, 1980; Lloyd, Webb & Primack, 1980; Sutherland, 1986; Primack, 1987; Stephenson, Devlin & Horton 1988) and a steadily increasing sophistication in the 'telling of adaptive stories' which purport to 'explain' observed behaviour (Gould & Lewontin, 1979).

The search for generalizations in the mass of observational data is exemplified in studies such as those of Sutherland (1986) who analysed fruit set values in 447 species, relating them to compatibility, breeding system, life form, latitude, type of fruit and type of pollination. One could envisage extending the list of species by orders of magnitude and the number of parameters to be considered many-fold, but it seems

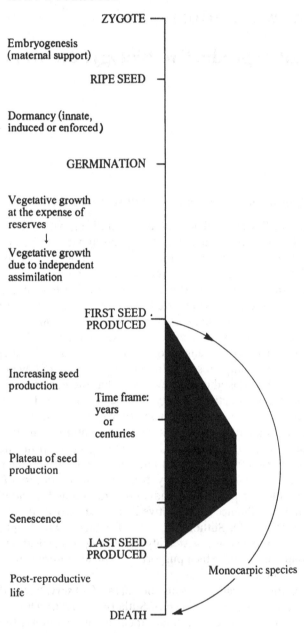

ZYGOTE

Embryogenesis
(maternal support)

RIPE SEED

Dormancy (innate,
induced or enforced)

GERMINATION

Vegetative growth
at the expense of
reserves
↓
Vegetative growth
due to independent
assimilation

FIRST SEED
PRODUCED

Increasing seed
production

Time frame:
years
  or
  centuries

Plateau of seed
production

Senescence

LAST SEED
PRODUCED

Post-reproductive
life

Monocarpic species

DEATH

Fig. 1. Diagram of the life cycle of a higher plant.

doubtful to the present author that there is any guarantee of significant generalizations emerging from this approach.

Reproductive effort

In an attempt to push the study of plant reproductive biology into a more quantitative and formal mould in a theoretical sense, several authors have sought to apply the ideas of MacArthur (1962) concerning r and K selection. At the heart of this approach is the notion that species or sub-specific taxa can be compared with respect to their allocation of resources to reproductive effort under density-dependent and density-independent conditions of mortality. Populations in habitats which give rise to high density-independent mortality (r-strategists) will be selectively favoured to allocate more of their resources to reproductive activities, which will prejudice their capacity to reproduce under crowded conditions. By contrast, populations in habitats which generate high density-dependent regulation (K-strategists) will be selectively favoured to allocate more of their resources to non-reproductive growth and development, at the expense of their capacity to reproduce in environments where there is a high level of density-independent mortality. Examples of this approach with plants are to be found in Gadgil & Solbrig (1972) for *Taraxacum*; Abrahamson & Gadgil (1973) on *Solidago*; and the series of papers on *Plantago* from Antonovics' group (cf. Primack & Antonovics, 1982). Studies such as these serve to provide some description of the ways in which populations respond under a variety of conditions. One intrinsic weakness of this approach is that it encourages the generation of comparative data and the extent to which these offer the prospect of any deeper understanding of the mechanisms involved tends to be ignored.

At the root of the practical problems which are soon encountered in applying theories of reproductive effort is the problem of definitions. Over the past twenty years ecologists and evolutionary biologists have approached the problem of reproductive effort from the standpoint that there must be a trade-off between the consumption of resources by an individual to meet its current reproductive activity and the use of resources for its own survival in order to achieve future reproduction. These notions lead in several directions: (i) intensive mathematical modelling by population geneticists to represent this trade-off in terms of fitness sets (see, for example, Stearns, 1976; Goodman, 1979; (ii) detailed considerations of the allocation of resources to the sub-sets of attributes of seed production, i.e. how many, how big and then how much for dispersal, dormancy, establishment, predator protection and environmental protection, etc. (for a review of this approach, see Foster, 1986).

Reproductive effort and the question of currency

Bazzaz & Reekie (1985), in a discerning review of work to date on reproductive effort in plants, express serious doubts as to whether the theoretical predictions alluded to above are borne out for a variety of plants in a variety of environments. They lay emphasis on the practicalities of measuring reproductive effort: as, for example, is biomass allocation to flowers and fruits a good indicator of allocation to reproduction; and what proportion of the respiration of a plant can one sensibly allocate to its reproductive effort? It is my view that this last little practical difficulty exemplifies a fundamental error in this whole approach. It may be technically convenient (Bazzaz, Carlson & Harper, 1979) or intellectually satisfying to suppose that one can use biomass or energy as a common currency for the costing of reproductive effort but there is absolutely no *a priori* reason why such a relationship should exist (Trivers, 1972; Dawkins, 1989). In many environments photosynthate is not limiting to growth; the elaboration of attributes of flowering response, floral development, embryogenesis, seed morphology, protection and dispersal mechanism can be amply provided for. The key elements in the development of this complex reproductive apparatus are the genes which control its developing and functioning. The task then becomes one of unravelling the genetic control of flowering and fruiting attributes in all their manifestations, a task further complicated by the fact that the genes involved do not act in isolation, they interact with one another.

New approaches to the study of plant reproductive biology

In suggesting the need for new approaches to plant reproductive biology I do not presume to advocate the abandonment of established lines of investigation; though I do imply that the slavish adherence of many ecologists to facile and inappropriate notions of reproductive effort could usefully be dropped.

If one produces haploid barley from anther culture, the plants that develop will produce an inflorescence but meiosis being blocked, there is no subsequent embryo development or filling of the grain. Under these circumstances axillary buds at the base of the stem grow out and the plant develops as a vegetative perennial. In *Poa annua*, races developing in hot dry microhabitats such as suburban streets or dry rock crevices produce a few tillers which carry an abundance of flowers and seeds and then die. From habitats such as well-watered lawns and cool damp montane locations, races of the same species have been described in which only a few of the tillers develop a weak inflorescence bearing few flowers and seeds, whereas other tillers adopt a horizontal habit, forming runners which root at the nodes and thereby promulgate the vegetative spread of the plant. The genetics of these race differences have yet to be determined but we may presume that, as in haploid barley, they relate to the genetic determinants of the relative sink strengths of the developing seeds and the axillary buds. In a recent review, Ho (1988) has discussed genetic factors influencing sink size and strengths, pointing to ways in which genes affecting the enzymes of starch biosynthesis and abscisic acid metabolism may influence these attributes (see also chapters by Ho and Duffus, this volume).

Recent studies with *Arabidopsis thaliana* (Meyerowitz, Smythe & Bowman, 1989) and *Antirrhinum majus* (Almeida *et al.*, 1989) have illustrated the successful application of transposon tagging to the isolation of genes controlling floral morphogenesis. It has also been established that transposon-mediated modification of promoter sequences of genes encoding enzymes on the anthocyanin biosynthetic pathway in *A. majus* can alter both the intensity of pigmentation and the pattern of pigment distribution over the surface of the petals (Martin *et al.* 1985). Studies such as these, and analogous work with mobile elements in bacteria and insects, carry implications of a role for these elements in the evolution of attributes of the reproductive system (Syvanen, 1984). Similar approaches are now being used with the Ac transposable elements from maize introduced into *Arabidopsis* for the isolation of genes involved in the vernalization response, timing of flowering and development of the siliqua. In these studies one sees the beginnings of combined molecular biological and genetic approaches to the study of the reproductive biology of plants. Further instances could be cited; for example, new work on gene expression in the ripening of fruits (Brady, 1987). It may be objected that this is a symposium of the Plant Environmental Biology Group of the SEB; but such an evasion will not do. My point is that both environmental and intrinsic developmental factors govern the expression of the genes affecting the reproductive biology of plants. The tools for identifying and measuring the activity of genes controlling reproductive behaviour are now developing rapidly; one hopes that this may lead to a new synthesis in which ecologists will align themselves with colleagues involved in these developments, for it is along this track that I believe that we shall come to a more comprehensive and convincing account of what constitutes the reproductive efforts of plants and a clearer view of the determinants of reproductive fitness.

References

Abrahamson, W.G. & Gadgil, M. (1973). Growth form and reproductive effort in Golden rods (*Solidago*, Compositae). *American Naturalist* **107**, 651–61.

Almeida, J., Carpenter, R., Robbins, T.P., Martin, C. & Coen, E.S. (1989). Genetic interactions underlying flower color patterns in *Antirrhinum majus*. *Genes and Development* **3**, 1758–67.

Baker, H.G. (1972). Seed weight in relation to environmental conditions in California. *Ecology* **53**, 998–1010.

Bazzaz, F.A., Carlson, R.W. & Harper, J.L. (1979). Contribution to reproductive effort by photosynthesis of flowers and fruits. *Nature* **279**, 554–5.

Bazzaz, F.A. & Reekie, E.G. (1985). Meaning and measurement of reproductive effort in plants. In *Studies in Plant Demography: A Festschrift for John L. Harper* (ed. J. White), pp. 373–87. London: Academic Press.

Brady, C.J. (1987). Fruit ripening. *Annual Review of Plant Physiology* **38**, 155–78.

Dawkins, R. (1989). *The Selfish Gene*. Oxford University Press.

Foster, S.A. (1986). On the adaptive value of large seeds for tropical moist forest trees: a review and synthesis. *Botanical Review* **52**, 260–99.

Gadgil, M. & Solbrig, O.T. (1972). The concept of r- and K-selection: evidence from wild flowers and some theoretical considerations. *American Naturalist* **106**, 14–31.

Gautier-Hion, A., Duplantier, J.-M., Quris, R., Foer, F., Sourd, C., Decoux, J.-P., Dubost, G., Emmons, L., Erard, C., Hecketweiler, P., Mourgazi, A., Roussilhon, C. & Thisllay, J.-M. (1985). Fruit characters as a basis of fruit choice and seed dispersal in a tropical forest vertebrate community. *Oecologia* **65**, 324–37.

Goodman, D. (1979). Regulating reproductive effort in a changing environment. *American Naturalist* **113**, 735–48.

Gould, S.J. & Lewontin, R.C. (1979). The spandrels of San Marco and the Panglossian paradigm: a critique of the adaptationist programme. *Proceedings of the Royal Society of London* B **205**, 581–98.

Heinrich, B. (1975). Energetics of pollination. *Annual Review of Ecology and Systematics* **6**, 139–70.

Ho, L.C. (1988). Metabolism and compartmentation of imported sugars in sink organs in relation to sink strength. *Annual Review of Plant Physiology and Plant Molecular Biology* **39**, 355–78.

Janzen, D.H. (1977). Why fruits rot, seeds mold and meat spoils. *American Naturalist* **111**, 691–713.

Lloyd, D.G. (1980). Sexual strategies in plants. I. An hypothesis of serial adjustment of maternal investment during one reproductive session. *New Phytologist* **86**, 69–79.

Lloyd, D.G., Webb, C.J. & Primack, R.B. (1980). Sexual strategies in plants. II. Data on the temporal regulation of maternal investment. *New Phytologist* **86**, 81–92.

MacArthur, R.H. (1962). Some generalised theorems of natural selection. *Proceedings of the National Academy of Sciences* **48**, 1893–7.

Martin, C., Carpenter, R., Somer, H., Saedler, H. & Coen, E.S. (1985). Molecular analysis of instability in flower pigmentation of *Antirrhinum majus* following isolation of the *pallida* locus by transposon tagging. *EMBO Journal* **4**, 1625–30.

Meyerowitz, E.M., Smythe, D.R. & Bowman, J.L. (1989). Abnormal flowers and pattern-formation in floral development. *Development* **106**, 209–17.

Mooney, H.A. (1972). The carbon balance of plants. *Annual Review of Ecology and Systematics* **3**, 315–46.

Primack, R.B. (1987). Relationships among flowers, fruits and seeds. *Annual Review of Ecology and Systematics* **18**, 409–30.

Primack, R.B. & Antonovics, J. (1982). Experimental ecological genetics in Plantago. VII. Reproductive effort in populations of *P. lanceolata* L. *Evolution* **36**, 742–52.

Stearns, S.C. (1976). Life-history tactics: a review of the ideas. *Quarterly Review of Biology* **51**, 3–47.

Stephenson, A.G., Devlin, B. & Horton, B.J. (1988). The effects of seed number and prior fruit dominance on the pattern of fruit production in *Cucurbita pepo* (zucchini squash). *Annals of Botany* **62**, 653–61.

Sutherland, S. (1986). Patterns of fruit set: what controls fruit:flower ratios in plants. *Evolution* **40**, 117–28.

Syvanen, M. (1984). The evolutionary implications of mobile genetic elements. *Annual Review of Genetics* **18**, 271–93.

Trivers, R.L. (1972). Parental investment and sexual selection. In *Sexual selection and the descent of man* (ed. B. Campbell), pp. 136–79. Chicago: Aldine.

R.F. LYNDON

The environmental control of reproductive development

Environmental control of flowering

The simple observation that many plants flower at a particular season or time of year implies that flowering is influenced and perhaps controlled by changes in the environment. All plants require conditions in which they can grow and develop in order to reproduce and so the ultimate environmental controls are those that determine plant distribution, and the main one may very often be temperature (Grace, 1987). But in all habitats, variations in the environment provide potential cues for the plant to make use of, so that the transition to reproductive growth coincides with the conditions most likely to lead to successful completion of flowering, fruiting and seed dispersal. Even in plants that do not respond to specific environmental changes, the onset and rate of progress of reproductive development will be determined by the general environment of the plant.

Are there any habitats in which the environment is constant and optimum so that growth and reproductive development are governed entirely by factors internal to the plant? The nearest approach to this is probably the tropical forest of SE Asia, which is virtually non-seasonal. However, even here flowering can be controlled by environmental changes, but because these are infrequent, mass flowering is also infrequent. When it does occur, up to 88% of the species may flower simultaneously (Appanah, 1985). The close correlation between an environmental change and subsequent flowering has been shown for the rain forest in Singapore, one of the least seasonal places in the world (Corlett, 1990). Two episodes of mass flowering were recorded a few months after an unprecedentedly dry, sunny month (Fig. 1). The two peaks of flowering (Fig. 1) also show that a single environmental event may subsequently result in flowering at different times for different species, presumably because, even in the same environment, flower development takes longer in some species than others. In other tropical forests

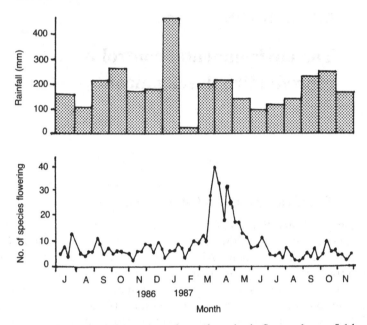

Fig. 1. Monthly rainfall and numbers of species in flower along a 5.1 km route through the forest in Singapore between July 1986 and November 1987. From March to May 1987, after a dry month (Feb. 1987), more than 150 species (out of a total of 787) flowered, with up to 40 species and 81 individuals flowering simultaneously. Normally in the otherwise uniform climate only about 6 species and 10 individuals were in flower at any one time. (After Corlett, 1990.)

the environment is less constant and flowering usually seems to be seasonal and to follow rainy or dry periods (Frankie, Baker & Opler, 1974). Mass flowering in tropical forests may occur only where the environment is normally very uniform and so when irregularities do occur many species respond.

Flowering at all times of the year may occur in cool temperate as well as in tropical environments. Some weeds and ornamentals can be in flower in the middle of winter in Edinburgh, Scotland (56° N) even after frosts if they have not been injured. Such plants can potentially flower all year round if they can grow, however slowly. The environment is controlling flowering only in so far as it allows or prevents growth. All-year flowerers in a very seasonal climate may be individuals that are either self-fertilized or apomictic (if wild plants) or ornamentals that do not depend upon the natural environment for cross-fertilization and propagation.

Why should plants use environmental cues to regulate flowering?

Plants may use environmental signals either to promote flowering, so that a favourable environment for reproductive growth can be quickly and successfully exploited, or to delay it until it can be achieved optimally as, for instance, in biennials. The critical parts of the reproductive process are pollination, fertilization and fruit and seed formation. The whole process has to be set in train, by the transition to flowering, several weeks or months in advance of eventual culmination. Not only must individuals be able to complete their reproductive development under favourable conditions but, for successful outcrossing, they must also flower synchronously with others of the same species. Those plants relying on specific animals or insects for dispersal may also require their seeds or fruits to reach ripeness at fairly precise times, which may depend on the timing of the previous stages of reproductive development.

If, however, the season is an abnormal one then the environmental changes, on which a plant may normally depend to promote flowering, may not occur. In this case the plant may need a fall-back position where internal controls can be overriding so that the plant reproduces, albeit suboptimally (Fig. 2). This would be consistent with the observation that most plants eventually flower, however unfavourable the conditions may be, so long as they can support some growth (Bernier, 1988) and with the view that, in the last resort, plant growth substances may provide a fail-

Fig. 2. Flowers of normal plants of *Silene coeli-rosa*, and of nutrient-deprived plants (grown in sand and watered only with distilled water). Without nutrients the flowers are very small, and only a single, small flower is formed instead of a many-flowered inflorescence (R.F. Lyndon, unpublished).

safe system to allocate scarce metabolic resources (Trewavas, 1986) and so allow flowering.

What environmental factors and cues are used by plants?

Any environmental variables are potentially available to plants for the control of flowering. The main variables are likely to be photoperiod, irradiance, temperature, and water availability. Mineral nutrients can also clearly regulate flowering in crop and ornamental plants, though this may be by a general effect on growth rather than a specific effect on flowering. An unusual case is that of *Nicotiana glutinosa*, in which high mineral supply causes the plant to behave as a LDP†, low supply as a SDP, and an intermediate supply as a DNP (Diomaiuto-Bonnand, 1974). For most plants in natural situations, minerals are not usually likely to change sufficiently and in a regular enough fashion to act to control flowering specifically, except perhaps in water plants, which may be subjected to seasonally varying concentrations of minerals. In tropical sea grasses of the Hydrocharitaceae, flowering can be induced by changes in salinity (Pieterse, 1985). Whether changes in mineral concentration are important for regulating the flowering of freshwater plants in the wild does not seem to have been studied, although the effects of minerals in controlling flowering of duckweeds (Lemnaceae) have been studied intensively in the laboratory (Kandeler, 1985). Freshwater aquatic plants, even submerged ones, that are rooted in the substratum, are unlikely to have their flowering controlled by changes in the mineral composition of the water because their main source of nutrients seems to be via their roots from the sediment rather than from the water itself (Hutchinson, 1975).

Photoperiod

Photoperiod as a variable controlling flowering has been widely investigated. It may delay, as well as promote, flowering. Because it is very precisely predictable within and between years, it is the most reliable environmental cue. Even on the equator it changes a little. In all latitudes, more than half the plants examined experimentally can respond to photoperiod by floral initiation. The difference between plants of the tropics and those of more temperate regions seems to be that short-day plants predominate in the tropics and long-day plants in higher latitudes.

†Abbreviations: SD, short day(s); SDP, short-day plant(s); LD, long day(s); LDP, long-day plant(s); DN, day-neutral; DNP, day-neutral plant(s); LSDP, long–short-day plant(s).

Table 1. *Photoperiodic regulation of floral initiation*

	Percentage that are		
	SDP	LDP	DNP
Temperate plants[a]	18	44	38
Tropical plants[b]	38	16	46

[a]Plants with a mainly non-tropical distribution.
[b]Plants with a mainly tropical distribution.
Source: Data condensed from Halevy (1985, 1989).

However, both long- and short-day types can be found at all latitudes (Table 1).

In some SDP of the tropics, measurement of daylength may be very precise. In *Corchorus* (jute) the critical photoperiod is 12 h and '. . even plants exposed to 12.25 h photoperiod failed to initiate flowers' (Nwoke, 1985*a*). In okra (*Hibiscus esculentus*) early varieties have a critical photoperiod of 12.5 h, and late varieties 12.25 h (Nwoke, 1985*b*).

Plants almost certainly measure photoperiod itself rather than whether it is changing, since '. . . there is little evidence to support the suggestion that plants might avoid the ambiguity [of the approximately equal equinoxial photoperiods of spring and autumn] by measuring "lengthening" versus "shortening" days, although it is known that some animals do . . .' (Vince-Prue, 1986). What is perhaps more likely is that the predominant effect is that of an increased number of cycles of a daylength progressively more favourable once the critical daylength has been passed. Most of the work on photoperiodism has been done with plants that react to a single cycle. But even in these (e.g. *Pharbitis*; Vince-Prue & Gressel, 1985), the critical daylength depends on number of cycles given and also, as in *Sinapis alba* (Table 2), on the age of the plant, which of course necessarily increases with an increasing number of cycles. In soybean, in which the number of flowering nodes also increases with the number of cycles given (Hamner, 1969), this is a function of the structure of the inflorescence and the sensitivity of the buds (Battey & Lyndon, 1990).

Table 2. *Induction of flowering in* Sinapis alba

Plant age (d)	% plants induced after . . . LD						
when LD given	1	2	3	4	5	6	7
15	0	10	15	30	50	75	90
30	50	100					
65	75	100					
90	100						

Source: From Bernier (1969).

Irradiance

A high enough irradiance is necessary to produce the photosynthate required for flowering (Bernier, Kinet & Sachs, 1981b). In some plants such as *Brunfelsia*, a plant of the forest understorey, flowering may be controlled by irradiance (Zimmer, 1985). Distinction may be made between a requirement, for flowering, of high irradiance or of high total light integral. The latter, as required by *Brassica campestris* (Friend, Deputy & Quedado, 1979), is more a measure of summed photosynthesis whereas the former is a requirement for a sufficiently high rate of photosynthesis.

Temperature

In crop plants with minimal or no photoperiodic requirements the time from sowing to anthesis and seed maturation can often be accounted for simply as a function of the accumulated day-degrees above a baseline value characteristic of that crop. This does not mean that light is unimportant for flowering but that temperature is broadly correlated with intercepted radiation (Milthorpe & Moorby, 1979). In general, increased temperature within the physiological range promotes flowering as it promotes growth, unless it is overridden by juvenility or by requirements for vernalization or photoperiod (see Table 4). However, flower initiation in LDP may often be promoted by lowering the temperature (Vince-Prue, 1975). Temperature controls flowering in many bulbs. Control is exerted at the stages of both flower initiation and flower development by increasing or decreasing the temperature (Rees, 1972).

Vernalization (by exposure to low temperatures in the range of about 0–12 °C) acts to promote progress toward flowering or to allow it to continue in low temperatures and short days. Characteristically, only

LDP or DNP respond to vernalization; SDP do not. However, 'many so-called "absolute" SDP . . . eventually produce flowers in LD or even in continuous light at 15° or below' (Bernier, Kinet & Sachs, 1981*a*). Examples are the SDP *Xanthium*, where low temperature (4 °C) would allow flowering 'even though the dark period was only 8.0 hours long' (Salisbury, 1969); and the SDP *Pharbitis*, in which, at 10 °C, flowering will proceed in LD (Ogawa, 1960).

In *Silene armeria*, high temperature, which promotes flower initiation, is perceived by the roots (Wellensiek, 1968). In some tropical trees, flowering can be promoted by keeping the roots cool (Leakey, Ferguson & Longman, 1981). On the other hand, in the SDP *Perilla* and *Pharbitis* and in the LDP *Blitum* it is apparently the leaves that detect the low temperature (Bernier *et al.*, 1981*a*). Only in plants requiring vernalization has it been shown that the shoot apex itself perceives the low temperature. The effects of temperature on flowering may therefore not necessarily be directly on the shoot apex but indirectly through effects on other parts of the plant. In some cases it may be the environmental factors to which the roots are exposed that could be more important. At the least we have to consider the plant as a whole and not get too concerned only with air temperature when soil temperature could be equally, or more, important. Furthermore, the temperature of the leaves may not necessarily be the same as the air temperature, especially for leaves and plants in the epiclimate near the ground (Grace, 1987). Although night temperature is often suggested as being more important than day temperature in affecting floral initiation this may be because much of the work has been done with SDP where the night period would be longer than the day. When allowance is made for the differing day and night lengths it has been shown for *Chrysanthemum* that it is the mean temperature (weighted for the different day and night lengths) that is important (Cockshull, 1985).

Interaction of temperature with photoperiod

An important point is that the seasonal cycles of photoperiod and temperature are not coincident. For a given length of photoperiod, temperatures in the autumn tend to be higher than in the spring (Fig. 3). This, together with the interactions of photoperiod and temperature, may be sufficient to remove any ambiguity that photoperiod alone might produce. In the tropics, where temperature changes may be slight, this ambiguity may also be removed by plants having a dual daylength requirement, as in LSDP, such as *Bryophyllum* (Zeevaart, 1985).

The interacting effects of temperature and photoperiod in controlling

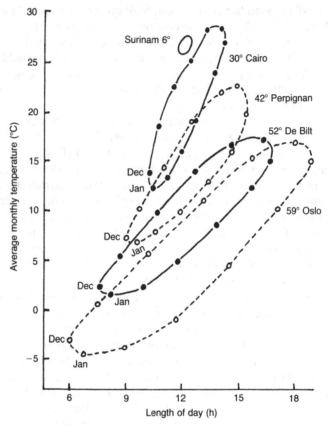

Fig. 3. Interaction of photoperiodic and temperature cycles. When temperature is shown as a function of daylength for each month, the resulting ellipsoid graphs describe the annual photothermal cycle characteristic of each latitude and locality. (After Ferguson, 1957.)

flowering pose the question of what their adaptive significance may be. Indeed in wheat, sensitivity to photoperiod may be a disadvantage, resulting in lower yields when environmental conditions are otherwise apparently optimum (Marshall *et al.*, 1989).

In general, flowering in SDP is reduced when night temperature is lowered (Vince-Prue, 1975), consistent with flowering being promoted by higher night temperatures. In LD plants, if the function of LD is simply to promote flower initiation, then we might similarly expect higher temperature to promote flowering and reduce the LD requirement. This is the case for the LDP *Rudbeckia*, *Calamintha* and *Silene*, but only at high temperatures (30 °C) (Bernier *et al.*, 1981*a*). However, this is the opposite of what is usually found. Much more common is the promotion of flower initiation in LDP by *lower* night temperature (see Vince-Prue, 1975; Bernier *et al.*, 1981*a*). Flower initiation in many LDP is therefore delayed by either SD or increased temperature, although subsequent development of the flower is often promoted by higher temperatures.

The interaction of temperature and photoperiod in both LD and SD is shown by a *Begonia* hybrid (which is essentially a SDP), in which higher temperatures promote flowering in SD but delay it in LD. Thus, although lower temperatures delay flowering in SD in this SDP, they promote flowering in LD (Heide, 1969). Consistent with these general trends, in the crop legumes, warm conditions (especially at night) can compensate for shorter nights in SDP and cool conditions can compensate for longer nights in LDP (Summerfield & Roberts, 1985). In general, in SDP the SD requirement is less stringent if temperatures are high, and in LDP the LD requirement is more stringent if temperatures are high.

How can we rationalize these effects of temperature and photoperiod? In SDP flowering is delayed by LD and by low temperatures; in LDP flowering is delayed by SD but not by low temperature. SDP can therefore be regarded as a strategy for delaying flowering by LD in high temperatures; and LDP a strategy for delaying flowering by SD in high temperatures but allowing it to go ahead irrespective of daylength at low temperatures, so that short days do not inhibit development in an otherwise permissive but non-optimum environment. SDP can therefore be thought of as a strategy to delay or synchronize flowering in an otherwise optimum environment for flowering; whereas LDP are a strategy to control and optimize flower development in a period of LD irrespective of the environment before the LD occur.

Water availability

Water availability as a factor controlling flowering is most obvious in habitats with seasonal rainfall. These may range from tropical forests to deserts but with the common factor of having high evapotranspiration rates so that water availability to the plant may be readily altered. Some plants respond to rain directly, others require a dry period first (see Singaporean forest, Fig. 1; *Tabebuia*, see p. 25). Coffee, for instance, requires water stress to break flower bud dormancy but then requires water uptake for subsequent bud development and flowering (Alvim, 1985).

Agave is a case where only very careful study over a decade has revealed that water availability is the controlling factor in flowering.

Agave plants are monocarpic, flowering when fully mature in size and about 50 years old. Flowering seemed to be determined only by the age and size of the plant but Nobel (1987) has shown that flowering of *Agave* in the Californian desert follows a period of summer rain two years previously. Small plants flowered only when attached by rhizomes to a large plant. Flowering years were followed by poor flowering years, irrespective of the rainfall two years previously, because most of the eligible large, old plants had already flowered the year before. Whether flowering of other monocarpic plants such as the 40-year bamboos and the talipot palm is also triggered by environmental factors when they are ripe to flower is not known. However it may be noted that synchronously flowering large types of bamboo do not occur within 5–10° of the equator (Janzen, 1985), suggesting that photoperiod could perhaps be implicated in triggering their flowering.

What distinguishes plants using different environmental controls?

Habitat and seasonality of the habitat

Plants as opportunists control flowering by using those seasonal factors of their environment that vary most or in a regular fashion. The particular environmental signals used by plants can therefore depend on their habitat. In the tropics the environment will tend to vary according to the wetness or dryness of season; in temperate zones according to the light–temperature regime. In deserts the overriding factor is clearly water availability, but different desert annuals flower at different times of the year. Desert annuals that flower in the late winter or early spring as days lengthen are facultative LDP. Summer annuals, flowering as days shorten in late summer or autumn, are facultative SDP. 'All annuals are facultative (quantitative) photo- and thermoperiodic plants' (Evanari & Gutterman, 1985). Desert perennials that are chamaephytes and nanochamaephytes are day-neutral (DN) and flower whenever water is available. Some all-year-round flowering 'bunch' grasses (*Astrebla*) of arid and semi-arid Australia are apparently DN for flower initiation but facultative SDP for inflorescence development. Other perennial 'bunch' grasses, which flower in spring or summer, may be facultative LDP. These desert plants show facultative responses to most environmental factors, including photoperiod, so that any time water becomes available they have the potential to flower, the timing of which can then be influenced by other environmental factors (Evanari & Gutterman, 1985).

In temperate regions, sand dunes are a comparable habitat in which water availability may vary greatly, and in which the annuals show facultative responses to other environmental factors. Here a further requirement for flowering may be vernalization, preventing premature winter flowering. In Britain, all species of dune annuals examined had a vernalization requirement for flower initiation and flowered best in long days and high temperatures. Most behaved as cold-requiring, facultative long-day plants (Pemadasa & Lovell, 1974).

Life form

The evolutionary origin of groups may predispose them to be of a particular response type, and so may the life form (Table 3), which may also be linked to taxonomic status; for example, Liliaceae are typically bulbs. The flowering of bulbs and corms is often controlled by temperature, or by water availability. This is not surprising since the initiation and early development of the leaves and flowers takes place in an underground organ that is normally inaccessible to light signals. (However, *Colchicum* corms 10 cm or more below the soil surface can react to photoperiod, the light apparently transmitted via the dead tissues linking the corm with the surface (Gutterman, 1989).)

Another life form is that of annuals and ephemerals. These must be able to flower in any year and so must have a flexible response to environmental signals. The facultative nature of the responses of desert and sand dune annuals has been discussed above. Similar facultative responses have been shown for eight annual species growing in a limestone community in England. All germinated in the autumn and flowered in the spring. A period at low temperatures of 5 °C or below during the winter was required for flower initiation and for normal flowering, which

Table 3. *Ways in which flowering can be controlled according to life form*

Herbs, annuals, ephemerals: Breaking of seed dormancy and then rapid life cycle. Promotion of flowering by environmental factors.

Biennials: Juvenility and/or cold or SD requirement, followed by promotion of flowering by photoperiod (usually LD) or adequate temperature.

Bulbs: Temperature or water; usually not accessible to other environmental factors until emergence, when photoperiod may affect floral development or regulate competition of flowering with bulbing.

Perennials: Breaking of dormancy, or continued growth, of preformed buds by temperature, photoperiod, or water stress.

was subsequently promoted by LD, so that these annuals, too, behaved as facultative LDP (Ratcliffe, 1961).

Other characteristic and common life forms are biennials, perennials, and trees. Biennials can have a response which is much more dependent on the environment because they are concerned with delaying or preventing flowering until specific conditions (often low temperatures followed by LD) have been met. Perennials and trees can have a much less specific and more general response to environmental factors, because they get many chances to flower and can afford not to flower for one or more seasons. Trees characteristically flower best at irregular intervals depending on favourable conditions. Perennials like the SDP *Chrysanthemum* and *Ribes* may require a fresh stimulus each year for flower induction. This implies that the plant as a whole is not induced, only the leaves. The leaves die or are abscised over the winter and so the whole process of floral induction has to start again the next year. A much studied example of leaf-only induction is *Perilla* (Zeevaart, 1969, 1986). In such plants, later-formed leaves can develop in a different photoperiod from the earlier leaves and so as the season progresses and the photoperiod becomes more unfavourable, the leaves then formed may increasingly override the effects of older induced, but senescing, leaves and so flowering can become inhibited after it has started.

All similar life forms in a given community may not necessarily have similar flowering responses. Different genotypes even within the same species may flower at slightly different, but distinct, times. We do not know to what extent there is, in any one plant community, a mixture of types showing different responses to the environment, e.g. a mixture of LDP, SDP, DNP, and plants with a juvenile period. Since all plants in a community do not usually flower simultaneously, there is presumably a mixture of response types or else different plants take different times for flower development to anthesis. Either could be a useful strategy to ensure that for species growing together, and exposed to essentially the same environmental signals, flowering is spread over a period, so perhaps making maximum use of insect pollinators.

Taxonomic position, genetics, and nucleotype

Response to environmental factors is often similar in closely related plants. Many plant families, such as the Compositae, encompass a range of photoperiodic responses although, within that family, closely related genera and species often (but not invariably) show the same response. In the grasses (Table 4) and the legumes (Summerfield & Roberts, 1985) photoperiodic and vernalization response is quite closely linked to taxo-

Table 4. *Control of flowering in grasses*

Festucoidae	
Festuceae	Species requiring
Hordeae	vernalization and/or long
Agrostideae	days are all in this
Aveneae	group
Phalarideae	
Stipeae	Mainly temperate distribution
Panicoideae	
Paniceae	Species requiring short days
Andropogoneae	are all in these groups
Maydeae	
Chloridoideae	No species in these groups
Chlorideae	are long-day plants
Zoysieae	or require vernalization
Eragrosteae, etc.	(*Cynodon* seems the only
	exception)
Oryzoideae	Mainly tropical and
Oryzeae	subtropical distribution

Source: After Evans (1964).

nomic relatedness within the family. On the other hand in some families there are genera, such as *Nicotiana*, which includes species with SD, LD and DN responses, which may be due to single gene differences (Lang, 1989). Even within a single species there may be different photoperiodic requirements shown by different ecotypes. *Themeda australis* is an Australian grass in which the populations from tropical latitudes are SDP and those from temperate latitudes are LDP, while the most southerly ecotypes also require vernalization (Evans, 1989).

The extent to which taxonomic status, and the related built-in constraints to respond to different environmental factors, determines plant distribution and flowering response has not been thoroughly explored. Plants in some families, such as the grasses (Table 4) and the legumes (Summerfield & Roberts, 1985) have sub-sections of closely related species that show similar environmental responses and have similar tropical or temperate distributions.

Genotype can produce different responses to environment within the same species, not only at different latitudes and locations but also within populations growing at the same location. Slight differences in genotype are presumably the explanation for plants of the same species, but from different provenances, flowering each year at different times but in the same order. The effect of genotype on flowering response has been particularly studied in wheat, in which single gene differences can affect response to vernalization (Law, Worland & Giorgi, 1976) and photoperiod (Law, 1987).

The effect of nucleotype on growth and latitudinal distribution of cereals has been noted by Bennett (1987). The nucleotype (essentially the amount of DNA in the nucleus) places restrictions on the life cycle by determining the minimum cell cycle time. If the amount of DNA per nucleus is too large then this does not allow the apex to go through a sufficient number of cell divisions in the season to allow flowering and fruiting to be completed and so the plant is necessarily a perennial. The length of time taken for meiosis is, in general, correlated with the nucleotype (Bennett, 1971). In seven cereal species, distribution was determined by nucleotype at the season when photoperiodic requirements for the species were not fulfilled (summer for the SDP and winter for the LDP) but the effect of nucleotype was overridden by photoperiod when the latter was optimal (winter for the SDP and summer for the LDP) (Fig. 3 in Bennett, 1987). These observations apply to where the crops are grown in summer or winter, not specifically to their flowering. The effect of nucleotype on flowering *per se*, although implicit in the cereal data, has not been examined explicitly for any species.

Control of flower development, gamete formation, and pollen release

The effects of environment can be on both flower initiation and subsequent development (Table 5). Where flowering is scored only by the appearance of visible flowers or buds, as it has often been in crop plants, especially the legumes, the data are sometimes insufficient to be sure what developmental stages are being affected most by environmental signals. For the plant this may perhaps be irrelevant, but for the researcher it is important to distinguish between the factors affecting flower initiation and flower development if we are to understand the processes involved in controlling flowering and the critical times at which the environment exerts its control.

In plants such as Feltham First peas (an early variety), the flowers are already formed in the seed and so any effects of the environment on flowering after germination are necessarily on flower development. But this also points to the possibility that in such plants the environment could have a significant control on flower initiation and early development

Table 5. *Potential control points in flowering*

Growth in general
Inhibition, e.g. by dormancy (as in deciduous trees), temperature, photoperiod, or by lack of rain. Promotion, so that the life cycle is completed in a shorter period, e.g. as in ephemerals, and becomes dependent only on internal factors such as nucleotype (q.v.).

Vegetative growth
Inhibition or promotion, so that onset of reproductive growth is favoured or retarded.

Flower initiation
Promotion or delay by photoperiod, temperature, juvenility, etc.

Flower development
Promotion by increased growth rate or release of bud dormancy; and control by determining sex ratio and timing of anthesis.

during the development and maturation of the seed. Possible effects of the environment in such early flower development do not seem to have been studied.

Flower development may be promoted by environmental conditions the same as, or different from, those affecting flower initiation (Vince-Prue, 1975; Kinet, Sachs & Bernier, 1985). In SD plants the requirement for SD may be more stringent for flower development than for flower initiation. Late flowering varieties of *Chrysanthemum*, for example, require '. . a shorter day for the development of an inflorescence than for its initiation, thus ensuring that normal inflorescences and open flowers occur only in the autumn' (Vince-Prue, 1986). In legumes 'the requirement for shorter or longer photoperiods . . . becomes progressively more stringent after the first flowers have been initiated . . . Successive stages of reproductive ontogeny may [also] have narrower temperature limits' (Summerfield & Roberts, 1985). Very often, if plants receive only the minimum photoperiodic requirement for flower induction this is insufficient to result in full flower development. Similarly, in LD plants sustained LD usually hasten flower development.

Flowering may be controlled by release of flower buds from dormancy rather than by their initiation *per se*. This is essentially what happens in the grass *Alopecurus* (a LDP) grown in high latitudes (Heide, 1986); the flowers are initiated in the autumn but the control 'to prevent precocious stem elongation and heading . . . is exerted by the combination of short days and low temperatures, which effectively halted flower development'. Winter may intervene between flower initiation and anthesis, as in deciduous trees, without any specific control perhaps being necessary to prevent flowering the first year. The interval between flower initiation and anthesis may occupy a relatively large proportion of the developmental cycle, as it does, for instance, in barley where the period from the beginning of ear initiation to anthesis is about two months, approximating to 40% or more of the period from sowing to seed harvest (Kirby, 1977). In perennials, if flowers are initiated late in the summer and the period between initiation and anthesis is normally long then inhibition of growth by the interpolation of winter during this period will necessarily delay flowering to the next season. However, there may be other more specific controls, such as the requirement for a minimum period at 15 °C for flower opening in apple (Goldwin, this volume). It is not clear whether photoperiod (SD) may prevent deciduous trees from flowering late in the season and ensure that they flower only the next year when days are sufficiently long. In many temperate and tropical trees, seasonal control of flowering may therefore be exerted by inhibition of growth and final flower development in unfavourable conditions rather than by a positive promoting effect. We really know relatively little about the control of flowering in many perennials. Environmental factors would also interact with genotypic and nucleotypic controls of flowering but the extent of this interaction has not been studied.

The structure of the flower can be controlled by environmental conditions. Sex expression can be controlled by the environment in a number of plants (Heslop-Harrison, 1972), such as *Spinacia* (Metzger & Zeevaart, 1985) and *Begonia* (Heide, 1969) in which LD promote femaleness. On the other hand, in the Cucurbitaceae femaleness tends to be promoted by SD (Rudich, 1985a,b) and in *Pinus*, SD promote female coning (Longman, 1982). A high proportion of tropical forest trees (30% or more) produce separate male and female flowers, and the majority tend to be dioecious (Longman, 1985) but whether these proportions are affected by environmental conditions is not known.

Environmental conditions may not only affect the presence or absence of male or female organs but may also cause deviations from the typical numbers of floral parts. Such meristic variation has been shown to be under environmental (temperature) control as, for example, in *Silene*, in which as many as one-third of the flowers could be affected (Lyndon, 1979). The floral parts which showed greatest variation in number were the stamens. The extent of meristic variation in wild populations, and whether it might be of any significance in causing small changes in the sex ratio, is not known.

When reproductive growth continues in environmental conditions that are unfavourable for it, for instance out of season or at the end of the season, the inflorescence and flowers may revert to vegetative growth and so show varying degrees of reversion. Vivipary is a form of reversion where either the inflorescence axis or the flower axis produces vegetative structures instead of flowers (Battey & Lyndon, 1990).

The reproductive functioning of the flower can also be controlled by the environment. Flower opening may depend on a sufficient light intensity, as in the Livingstone Daisy (*Mesembryanthemum*), which opens only in strong sunlight. In *Moenchia* the flowers open in strong light, thus allowing cross-pollination, but in dull light the flowers remain closed and so self-pollination is favoured (Clapham, Tutin & Warburg, 1962). Cleistogamy, too, can be controlled environmentally as shown in *Lamium*. Flowers are all cleistogamous in SD but up to 30% open when the plants are grown in LD (Lord, 1979). The precise time of flower opening (anthesis) in *Pharbitis* is controlled by photoperiod and temperature so that the flowers open just before dawn. The higher the temperature, the later is flower opening (Kaihara & Takimoto, 1979). Anthesis in *Tabebuia* may be controlled by water availability. *Tabebuia* trees in tropical rain forest may flower synchronously, within a day or two, even though the individual trees can be kilometres away from each other. This is because they have reacted identically to a common stimulus, which appears to be heavy rain relieving water stress after a period of drought. Cell enlargement then causes opening of the flower buds (Borchert, 1986). Other species growing in the same locality responding to the same environmental signals, but on different developmental time scales, would flower at different times.

Pollen release in grasses can occur at very precise times of day, which are different for different species (Evans, 1964). The avocado (*Persea*) is unique in that the flowers open twice, on successive days. The first day the pistil is receptive; the second day the pollen is shed. This timing can be reversed by cool weather (Bergh, 1986). Environmental variables can therefore affect not only the initiation and growth of flowers, but also influence the sex ratio and the success of cross-pollination.

How do the effects of the environment interact with the plant to control flowering?

Environmental factors affect the plant in three main ways: through water status, changes in assimilate and mineral availability and distribution within the plant, and changes in the production of, and reaction to, endogenous plant growth substances. Water status in most plants will have a general effect in that if water is insufficient, growth will be affected and flowering may be retarded or impaired. In some plants, noted earlier, changes in water status may be an essential signal to provoke flowering or to synchronise it. Flowering may also depend on the rate and amount of assimilate production, as in *Brassica campestris* cv. Ceres (Friend *et al.*, 1979), or on changes in assimilate distribution (Sachs, 1987). However, alteration of assimilate production and distribution and increase of sucrose in the apical meristem during floral induction, as in *Sinapis* (Bodson & Outlaw, 1985) may, in many plants, be more a correlate of the transition to flowering rather than its cause (Bodson & Remacle, 1987).

The effects of the environment on flowering are probably transduced in the plant by the involvement of plant growth substances. Environmental changes have been shown to affect plant growth substance content in plants undergoing floral induction; also, applied growth substances can induce flowering in many plants (Zeevaart, 1978). Although gibberellins are effective in inducing flowering in many LD plants, and some conifers, there are no really reliable data to show that endogenous gibberellin content changes in a way closely correlated with flowering. The difficulties include knowing which part of the plant to measure and what magnitude of changes should be expected, and to what extent changes in tissue sensitivity may be important. Also, if a single substance or group of substances is not the sole cause of induction, then the meaning of any correlations or lack of them could be difficult to interpret. Difficulties are exemplified by the changes in the cytokinin content on flowering, which increases in leaves and root exudates of the LD plant *Sinapis* (Lejeune, Kinet & Bernier, 1988), and decreases in the SD plant *Xanthium* (Wareing *et al.*, 1977). Although correlated with flowering, application of cytokinin does not induce flowering in either of these plants and flowering in *Sinapis* could be induced by a LD even when the roots had been removed (Lejeune *et al.*, 1988).

The growth and development of excised flowers cultured *in vitro* can be affected by applied growth substances, and there is also evidence that floral organs are sources of growth substances (Table 6). Whether growth substances produced by one floral whorl influence and control the growth of successive whorls is not known, nor is it known to what extent the environment may affect flower form through an effect on endogenous growth substances. Reversion in *Impatiens* can best be interpreted as showing that photoperiod controls the formation, by the leaves, of substances which then move to the shoot apex where they directly control flower formation and flower form (Battey & Lyndon, 1990). In most plants (which do not show flower reversion) the direct effects of the environment on flower development may be very much less.

The plant in its normal growth tends towards flowering as the culmina-

Table 6. *Growth substances and flower development*

Floral organ	Growth substances produced	Growth substances promoting growth
Sepals	GA	GA, Auxin
Petals	GA	GA, Auxin
Stamens	Auxin, GA	GA, Auxin
Ovary	Auxin, GA	GA, Auxin
Whole flower	—	Cytokinin

Source: After Kinet, Sachs & Bernier (1985).

tion of the growth process. One point of view is that endogenous changes in growth substances will normally tend to favour the completion of flowering by promoting and coordinating normal development but that the environment can override this internal control, probably by upsetting the growth substance balance in the plant, to delay flowering until the plant has made sufficient growth to support it or until environmental conditions again become permissive. This suggestion, that growth substances are the normal coordinators of flower development, is similar to the view that growth substances act to coordinate growth by controlling resource allocation under poor growth conditions (Trewavas, 1986). One implication would be that under prolonged extreme conditions which, none the less, allow some growth, the normal growth substance balance may eventually be restored so that the plant then flowers despite unfavourable conditions.

Environmental control of flowering in wild plants

Many investigations of the effects of the environment on the control of flowering have been done on plants of agricultural or horticultural interest. Some have been done on wild plants that are common weeds. Crop plants may be suitable experimental material for demonstrating the effects of the environment on flowering because they are not necessarily specifically adapted to their habitat and so may be grown and experimented upon in what would, for wild plants, be unusual conditions. However, experiments on such plants may not be particularly relevant to plants growing in the wild in natural communities and subject to competition. How the environment controls reproductive behaviour in wild plants in most natural communities is simply not known and is an obvious subject for future research.

Acknowledgement

I am very grateful to John Dale for his helpful comments and suggestions.

References

Alvim, P. de T. (1985). *Coffea*. In *Handbook of Flowering*, vol. 2 (ed. A.H. Halevy), pp. 308–16. Boca Raton: CRC Press.

Appanah, S. (1985). General flowering in the climax rain forests of South-east Asia. *Journal of Tropical Ecology* 1, 225–40.

Battey, N.H. & Lyndon, R.F. (1990). Reversion of flowering. *Botanical Reviews* 56, 162–89.

Bennett, M.D. (1971). The duration of meiosis. *Proceedings of the Royal Society of London* B178, 277–99.

Bennett, M.D. (1987). Variation in genomic form in plants and its ecological implications. *New Phytologist* 106 (suppl.), 177–200.

Bergh, B.O. (1986). *Persea americana*. In *Handbook of Flowering*, vol. 5 (ed. A.H. Halevy), pp. 253–68. Boca Raton: CRC Press.

Bernier, G. (1969). *Sinapis alba* L. In *The Induction of Flowering: Some Case Histories* (ed. L.T. Evans), pp. 305–27. Melbourne: Macmillan.

Bernier, G. (1988). The control of floral evocation and morphogenesis. *Annual Review of Plant Physiology and Plant Molecular Biology* 39, 175–219.

Bernier, G., Kinet, J.-M. & Sachs, R.M. (1981a). *The Physiology of Flowering*, vol. 1. Boca Raton: CRC Press.

Bernier, G., Kinet, J.-M. & Sachs, R.M. (1981b). *The Physiology of Flowering*, vol. 2. Boca Raton: CRC Press.

Bodson, M. & Outlaw, W.H. (1985). Elevation in the sucrose content of the shoot apical meristem of *Sinapis alba* at floral evocation. *Plant Physiology* 79, 420–4.

Bodson, M. & Remacle, B. (1987). Distribution of assimilates from various source-leaves during the floral transition of *Sinapis alba* L. In *Manipulation of Flowering* (ed. J. Atherton), pp. 341–50. London: Butterworth.

Borchert, R. (1986). *Tabebuia*. In *Handbook of Flowering*, vol. 5 (ed. A.H. Halevy), pp. 345–52. Boca Raton: CRC Press.

Clapham, A.R., Tutin, T.G. & Warburg, E.F. (1962). *Flora of the British Isles*. 2nd edn. Cambridge University Press.

Cockshull, K.E. (1985). *Chrysanthemum morifolium*. In *Handbook of Flowering*, vol. 2 (ed. A.H. Halevy), pp. 238–57. Boca Raton: CRC Press.

Corlett, R.T. (1990). Flora and reproductive phenology of the rain forest at Bukit Timah, Singapore. *Journal of Tropical Ecology* 6, 55–63.

Diomaiuto-Bonnand, J. (1974). Variations du comportement photopériodique du *Nicotiana glutinosa* L. en fonction de la fréquence de l'apport en substances minérales au cours des arrosages. *Comptes Rendues de l'Academie des Sciences (Paris)* **278** D, 49–52.

Evanari, M. & Gutterman, Y. (1985). Desert plants. In *Handbook of Flowering*, vol. 1 (ed. A.H. Halevy), pp. 41–59. Boca Raton: CRC Press.

Evans, L.T. (1964). Reproduction. In *Grasses and Grasslands* (ed. C. Barnard), pp. 126–53. London: Macmillan.

Evans, L.T. (1989). *Themeda australis*. In *Handbook of Flowering*, vol. 6 (ed. A.H. Halevy), pp. 598–600. Boca Raton: CRC Press.

Ferguson, J.H.A. (1957). Photothermographs, a tool for climate studies in relation to the ecology of vegetable varieties. *Euphytica* **6**, 97–105.

Frankie, G.W., Baker, H.G. & Opler, P.A. (1974). Comparative phenological studies of trees in tropical wet and dry forests in the lowlands of Costa Rica. *Journal of Ecology* **62**, 881–919.

Friend, D.J.C., Deputy, J. and Quedado, R. (1979). Photosynthetic and photomorphogenetic effects of high photon flux densities on the flowering of two long-day plants, *Anagallis arvensis* and *Brassica campestris*. In *Photosynthesis and Plant Development* (ed. R. Marcelle, H. Clijsters and M. Van Poucke), pp. 59–72. The Hague: Junk.

Grace, J. (1987). Climate tolerance and plant distribution. *New Phytologist* **106** (suppl.), 113–30.

Gutterman, Y. (1989). *Colchicum tunicatum*. In *Handbook of Flowering*, vol. 6 (ed. A.H. Halevy), pp. 234–42. Boca Raton: CRC Press.

Halevy, A.H. (1985; 1989). *Handbook of Flowering*, vols 1–6. Boca Raton: CRC Press.

Hamner, K.C. (1969). *Glycine max* (L.) Merill. In *The Induction of Flowering: Some Case Histories* (ed. L.T. Evans), pp. 62–89. Melbourne: Macmillan.

Heide, O.M. (1969). Environmental control of sex expression in *Begonia*. *Zeitschrift für Pflanzenphysiologie* **61**, 279–85.

Heide, O.M. (1986). Primary and secondary induction requirements for flowering in *Alopecurus pratensis*. *Physiologia Plantarum* **66**, 251–6.

Heslop-Harrison, J. (1972). Sexuality of angiosperms. In *Plant Physiology. A Treatise*, vol. 6C (ed. F.C. Steward), pp. 133–289. New York: Academic Press.

Hutchinson, G.E. (1975). *A Treatise on Limnology*, vol. 3. New York: Wiley.

Janzen, D.H. (1985). Bambusoideae. In *Handbook of Flowering*, vol. 2 (ed. A.H. Halevy), pp. 1–3. Boca Raton: CRC Press.

Kaihara, S. & Takimoto, A. (1979). Environmental factors controlling the time of flower-opening in *Pharbitis nil*. *Plant and Cell Physiology* **20**, 1659–66.

Kandeler, R. (1985). Lemnaceae. In *Handbook of Flowering*, vol. 3 (ed. A.H. Halevy), pp. 251–79. Boca Raton: CRC Press.

Kinet, J.-M., Sachs, R.M. & Bernier, G. (1985). *The Physiology of Flowering*, vol. 3. Boca Raton: CRC Press.

Kirby, E.J.M. (1977). The growth of the shoot apex and the apical dome of barley during ear initiation. *Annals of Botany* 41, 1297–308.

Lang, A. (1989). *Nicotiana*. In *Handbook of Flowering*, vol. 6 (ed. A.H. Halevy), pp. 427–83. Boca Raton: CRC Press.

Law, C.N. (1987). The genetic control of day-length response in wheat. In *Manipulation of Flowering* (ed. J. Atherton), pp. 225–40. London: Butterworth.

Law, C.N., Worland, A.J. & Giorgi, B. (1976). The genetic control of ear-emergence time by chromosomes 5A and 5D of wheat. *Heredity* 36, 49–58.

Leakey, R.R.B., Ferguson, N.R. & Longman, K.A. (1981). Precocious flowering and reproductive biology of *Triplochiton scleroxylon* K. Schum. *Commonwealth Forestry Review* 60, 118–26.

Lejeune, P., Kinet, J.-M. & Bernier, G. (1988). Cytokinin fluxes during floral induction in the long day plant *Sinapis alba* L. *Plant Physiology* 86, 1095–8.

Longman, K.A. (1982). Effects of gibberellin, clone and environment on cone initiation, shoot growth and branching in *Pinus contorta*. *Annals of Botany* 50, 247–57.

Longman, K.A. (1985). Tropical forest trees. In *Handbook of Flowering*, vol. 1 (ed. A.H. Halevy), pp. 23–39. Boca Raton: CRC Press.

Lord, E.M. (1979). Physiological controls on the production of cleistogamous and chasmogamous flowers in *Lamium amplexicaule* L. *Annals of Botany* 44, 757–66.

Lyndon, R.F. (1979). Aberrations in flower development in *Silene*. *Canadian Journal of Botany* 57, 233–5.

Marshall, L., Busch, R., Cholick, F., Edwards, I. & Frohberg, R. (1989). Agronomic performance of spring wheat isolines differing for daylength response. *Crop Science* 29, 752–7.

Metzger, T.D. & Zeevaart, J.A.D. (1985). *Spinacia oleracea*. In *Handbook of Flowering*, vol. 4 (ed. A.H. Halevy), pp. 384–92. Boca Raton: CRC Press.

Milthorpe, F.L. & Moorby, J. (1979). *An Introduction to Crop Physiology*. 2nd edn. Cambridge University Press.

Nobel, P.S. (1987). Water relations and plant size aspects of flowering for *Agave deserti*. *Botanical Gazette* 148, 79–84.

Nwoke, F.I.O. (1985a). *Corchorus capsularis* and *C. olitorus*. In *Handbook of Flowering*, vol. 2 (ed. A.H. Halevy), pp. 324–30. Boca Raton: CRC Press.

Nwoke, F.I.O. (1985b). *Hibiscus esculentus*. In *Handbook of Flowering*, vol. 3 (ed. A.H. Halevy), pp. 140–9. Boca Raton: CRC Press.

Ogawa, Y. (1960). Über die Auslösung der Blütenbildung von *Pharbitis nil* durch niedere Temperatur. *Botanical Magazine Tokyo* **73**, 334–5.

Pemadasa, M.A. & Lovell, P.H. (1974). Factors controlling the flowering time of some dune annuals. *Journal of Ecology* **62**, 869–80.

Pieterse, A.H. (1985). Hydrocharitaceae (Sea Grasses). In *Handbook of Flowering*, vol. 3 (ed. A.H. Halevy), pp. 181–2. Boca Raton: CRC Press.

Ratcliffe, D. (1961). Adaptation to habitat in a group of annual plants. *Journal of Ecology* **49**, 187–203.

Rees, A.R. (1972). *The Growth of Bulbs*. London: Academic Press.

Rudich, J. (1985*a*). *Cucumis melo*. In *Handbook of Flowering*, vol. 4 (ed. A.H. Halevy), pp. 360–4. Boca Raton: CRC Press.

Rudich, J. (1985*b*). *Cucumis sativus*. In *Handbook of Flowering*, vol. 4 (ed. A.H. Halevy), pp. 365–74. Boca Raton: CRC Press.

Sachs, R.M. (1987). Roles of photosynthesis and assimilate partitioning in flower initiation. In *Manipulation of Flowering* (ed. J. Atherton), pp. 317–40. London: Butterworth.

Salisbury, F.B. (1969). *Xanthium strumarium*. In *The Induction of Flowering: Some Case Histories* (ed. L.T. Evans), pp. 14–61. Melbourne: Macmillan.

Summerfield, R.J. & Roberts, E.H. (1985). Grain legume species of significant importance in world agriculture. In *Handbook of Flowering*, vol. 1 (ed. A.H. Halevy), pp. 61–73. Boca Raton: CRC Press.

Trewavas, A.J. (1986). Resource allocation under poor growth conditions. A major role for growth substances in developmental plasticity. In *Plasticity in Plants* (ed. D.H. Jennings & A.J. Trewavas) (*Symposia of the Society for Experimental Biology* **40**), pp. 31–76. Cambridge University Press.

Vince-Prue, D. (1975). *Photoperiodism in Plants*. London: McGraw-Hill.

Vince-Prue, D. (1986). The duration of light and photoperiodic responses. In *Photomorphogenesis in Plants* (ed. R.E. Kendrick & G.H.M. Kronenberg), pp. 269–305. Dordrecht: Martinus Nijhoff.

Vince-Prue, D. & Gressel, J. (1985). *Pharbitis nil*. In *Handbook of Flowering*, vol. 4 (ed. A.H. Halevy), pp. 47–81. Boca Raton: CRC Press.

Wareing, P.F., Horgan, R., Henson, I.E. & Davis, W. (1977). Cytokinin relations in the whole plant. In *Plant Growth Regulation* (ed. P.E. Pilet), pp. 147–53. Berlin: Springer.

Wellensiek, S.J. (1968). Floral induction through the roots of *Silene armeria* L. *Acta Botanica Neerlandica* **17**, 5–8.

Zeevaart, J.A.D. (1969). *Perilla*. In *The Induction of Flowering: Some Case Histories* (ed. L.T. Evans), pp. 116–55. Melbourne: Macmillan.

Zeevaart, J.A.D. (1978). Phytohormones and flower formation. In *Phytohormones and Related Compounds – A Comprehensive Treatise,* vol. 2 (ed. D.S. Letham, P.B. Goodwin & T.J.V. Higgins), pp. 291–327. Amsterdam: Elsevier.

Zeevaart, J.A.D. (1985). *Bryophyllum*. In *Handbook of Flowering*, vol. 2 (ed. A.H. Halevy), pp. 89–100. Boca Raton: CRC Press.

Zeevaart, J.A.D. (1986). *Perilla*. In *Handbook of Flowering*, vol. 5 (ed. A.H. Halevy), pp. 239–52. Boca Raton: CRC Press.

Zimmer, K. (1985). *Brunfelsia pauciflora* var. *calycina*. In *Handbook of Flowering*, vol. 2 (ed. A.H. Halevy), pp. 85–8. Boca Raton: CRC Press.

S. J. OWENS

Pollination and fertilization in higher plants

Introduction

This paper focuses on events post-pollination, and will only describe the forms of pollen, stigma, style, ovary, and embryo sac where structure is relevant to the process being discussed. Data have been included on a selective basis; the paper, therefore, must not be regarded as an exhaustive analysis. For further information, a number of reviews have recently appeared relating to pollination and fertilization. These include reviews of the male gametophyte (Mascarenhas, 1989), pollen, pistil and reproductive function (Knox & Williams, 1986), pollen germination and tube growth (Heslop-Harrison, 1987; Steer & Steer, 1989), micro- and megasporogenesis and fertilization (Fougere-Rifot, 1987), the cytological basis of the plastid inheritance in angiosperms (Hagemann & Schroder, 1989), and molecular aspects of fertilization (Cornish, Anderson & Clarke, 1988). Self-incompatibility has also been deliberately given a low priority in this paper because excellent reviews have been published (Gibbs, 1986; Heslop-Harrison, 1983; Clarke et al., 1989) and in order to emphasize other aspects of the process of fertilization and its control. Where relevant in any discussion, self-incompatibility is briefly considered.

The pollen–stigma interaction

Stigma receptivity

The development of receptivity of the stigmatic surface may vary widely between species from several days prior to anthesis to several days after the flower opens (Palser, Rouse & Williams, 1989). In 'wet' stigmas receptivity often coincides with accumulation of stigmatic secretion and in most stigmas with a positive response for non-specific esterase (Heslop-Harrison and Shivanna, 1977). In protogynous and protandrous species stigma receptivity is strictly related to anther dehiscence (Richards,

1986). Some responses appear very specialized. In *Talinum mengesii* (Portulacaceae) (Murdy & Carter, 1987), for example, pollen germination is delayed for up to 2 h after compatible pollination. The delay depends entirely on the onset of stigma receptivity. Murdy & Carter (1987) consider that this trait promotes accumulation of pollen on the stigma and simultaneous pollen germination, which in turn may intensify male gametophyte competition. It also may encourage pollen to be deposited from more than one plant on a stigma, a situation which has been shown to be positively correlated with seed mass per fruit (Marshall & Ellstrand, 1986) and the quality of sporophytic offspring (Van der Kloet, 1984), although such a correlation appears not to be universal (Schemske & Pautler, 1984; Bertin, 1986). There is no published case of pollen touch-down initiating stigma receptivity; it appears to be a pre-programmed response.

Adherence, hydration, recognition

A number of common events have been identified following pollen touch-down on the stigma in a compatible pollination. Pollen adherence (particularly necessary for 'dry' stigma forms), pollen hydration (more accurately rehydration, since pollen grains lose water prior to anther dehiscence (Heslop-Harrison, 1987)) and pollen recognition (particularly in self-incompatible species) are characteristics that take place before pollen germination (Heslop-Harrison, 1987). There is some circumstantial evidence to indicate a sequential process (pollen adherence, for example, may be a prerequisite for pollen rehydration) and for the stigma to possess the machinery for ensuring a carefully regulated rehydration of pollen grains before their germination (Heslop-Harrison, 1979a,b).

Pollen germination

Pollen germination is temperature-dependent. An extreme example is found in *Juglans regia* or *Juglans nigra* cultivars where no pollen germinates on stigmas below 14 °C or at 40 °C (Luza, Polito & Weinbaum, 1987). Pollen germination rates are, however, very variable between species (Table 1). Grass pollen appears to germinate most rapidly *in vivo* and pollen of Orchidaceae most slowly. Although a considerable body of information is available on stigmatic surface secretions (Knox, 1984; Dumas *et al.*, 1988) there is no adequate explanation for such variation. There are two possible candidates, however. 'Dry' stigmas appear to be associated with more rapid germination, although data are very limited (Table 1), and trinucleate grains appear to germinate more rapidly than

Table 1. *Time to pollen germination and its relation with stigma form[a]*

	Time to pollen germination	Stigma type	Reference
Aloaceae			
Gasteria verrucosa	30 min	wet	Willemse and Franssen-Verheijen, 1988
Gramineae			
Secale cereale	75 s	dry	Heslop-Harrison, 1979*a*, *b*
Orchidaceae			
Oncidium	72 h	wet	Clifford & Owens, 1988
Cruciferae			
Brassica oleracea	60–180 min	dry	Roberts *et al.*, 1980; Ferrari *et al.*, 1985
Ericaceae			
Rhododendron nuttallii	60–120 min	wet	Palser *et al.*, 1989
Juglandaceae			
Carya illinoensis	180 min	wet	Wetzstein & Sparks, 1989
Leguminosae			
Trifolium pratense	10–20 min	wet	Heslop-Harrison & Heslop-Harrison, 1983
Bauhinia pentandra	180 min	wet	Owens, 1989
Bauhinia tomentosa	60 min	wet	Owens, 1989
Caesalpinia gillesii	30 min	wet	Owens, 1989
Caesalpinia pulcherrima	43 min	wet	Owens, 1989
Cercis chinensis	60 min	wet	Owens, 1989
Cercis siliquastrum	55 min	wet	Owens, 1989
Senna floribunda	20 min	wet	Owens, 1989
Oenotheraceae			
Oenothera organensis	<40 min	wet	Dickinson & Lawson, 1975
Myrtaceae			
Eucalyptus woodwardii	ca 48 h	wet	Sedgely & Smith, 1989
Portulacaceae			
Talinum mengesii	<30 min	—	Murdy & Carter, 1987
Proteaceae			
Grevillea banksii	>60 min	wet	Herscovitch & Martin, 1989
Solanaceae			
Lycopersicon peruvianum	ca 210 min	wet	Cresti *et al.*, 1977
Salpiglossis sinuata	30 min	wet	Hepher & Boulter, 1987

[a]As defined by Heslop-Harrison (1981).

binucleate (Hoekstra, 1979; Hoekstra & Bruinsma, 1979). The grasses are surely a special case: their pollens are often very short-lived (measured in minutes rather than hours) and their stigmas remain receptive for short periods, perhaps not more than 3–4 h (Heslop-Harrison, 1979b). Using anhydrous fixation techniques, Dickinson & Elleman (1985) have shown that the pollen grain coat and the stigma papillar surface appear adapted for strict control of water entry into the grain. In addition, different self-incompatibility (S) genotypes have been shown to have significant differences in stigma cuticle thickness, which may itself influence the rate of pollen hydration (Elleman et al., 1988).

It should not be surprising that in both self-incompatible and self-compatible species not all pollen germinates on the stigma or that some tubes abort on the stigma (Palser et al., 1989) or in the style. Both the genotype of the pollen and the environment may affect pollen viability and preclude germination or reduce and slow tube growth (Thomson, 1989).

Stylar penetration

Pollen germination is followed by pollen tube growth. In species with 'dry' stigma forms (e.g. Crocus (Heslop-Harrison & Heslop-Harrison, 1975) or Brassica oleracea (Roberts et al., 1980; Elleman et al., 1988)), penetration of the stigmatic cuticle is necessary before the pollen tubes can enter the style and grow towards the ovary, whereas in many other species the cuticle is ruptured during the normal development of the receptive stigma (Dickinson & Lawson, 1975) or during the act of insect visitation and pollination (e.g. Papilionoideae–Leguminosae) (Lord & Heslop-Harrison, 1984; Kreitner & Sorensen, 1984; Owens, 1985).

Post-pollination stigmatic responses

The stigma in Acacia species is cup-shaped and contains a drop of secretion, which is considered to aid pollen adhesion (Kenrick & Knox, 1981). Pollination, however, promotes the development of a massive post-pollination exudate within approximately 30 min of touch-down, regardless of the pollen's origin. Kenrick & Knox (1981) postulate that this is the result of a general pollination signal, which is initiated at the commencement of the pollen–stigma interaction.

Post-pollination responses are readily observable in the flowers of Orchidaceae. Stigma closure 24 h after either a self- or cross-pollination, and column and ovary swelling, are characteristic for the many members of the family (Clifford & Owens, 1988). There is considerable evidence to suggest that such responses are promoted by auxin; auxin has very recently been found in pollinia of orchids by using thin-layer chromatography techniques (S.C. Clifford & S.J. Owens, unpublished results).

Pollination also shortens the time to the onset of flower wilting, petal colour change, and abscission (Arditti, 1979; Lovell, Lovell & Nichols, 1987).

Pollen tube growth

Pre-programmed development of the style

A progressive, programmed development of the stylar transmitting tissue, which is known to be independent of pollination, has been found even though the activation of the pistil, as a consequence of pollination, is a well-known fact (Mulcahy & Mulcahy, 1988) and is dealt with later in this paper. Dickinson, Moriarty & Lawson (1982), for example, found that after the start of stigmatic activity (papillar cell elongation, cuticle rupture and secretion development) and before pollination, stylar canal cells developed in a 'wave' starting at the top of the pistil and continuing down the style. The stylar canal-cell cytoplasm became stratified, with nuclei, plastids and lipid droplets accumulating at sites away from the wall lining the canal whereas ER, dictyosomes and some mitochondria moved towards the canal wall. The canal lining wall itself developed transfer-cell-like features. Recently, S.J. Owens (1989 and unpublished data) has shown that the breakdown of the intercellular connections of cells surrounding the stylar canal of *Caesalpinia pulcherrima* are part of a pre-programmed, post-anthesis development and are not stimulated by pollination.

Pollen tube extension

The pollen tube extends by tip growth (Steer & Steer, 1989); pollen tube growth may be through a stylar canal, intercellularly through a core of transmitting tissue or through an area of cellular debris, which appears more akin to growth in a stylar canal. In Liliaceae (Dickinson *et al.*, 1982; Willemse & Franssen-Verheijen, 1988) and Leguminosae (Heslop-Harrison, 1982), pollen tube growth initially is through solid or semi-solid transmitting tissue, and subsequently through a stylar canal.

The pollen tube has recently been the subject of major reviews (Heslop-Harrison, 1987; Steer & Steer, 1989), which have examined the current data on the tube cytoplasm, wall structure and the interaction of pollen tubes with calcium ions. Pollen tubes have a marked cytoplasmic polarity; distinct zones can be recognized in the cytoplasm based on their complements of nuclei and cells and cytoplasmic organelles. Dictyosomes are particularly important in tip growth: the vesicles supply both wall material and plasma membrane components (Steer & Steer, 1989). The tube wall comprises an outer pectic layer and an inner callosic–cellulosic layer, which is absent from the tube tip. The highest concentrations of calcium (both membrane-bound and free) are found at the tube tip. Some of the most interesting data concern the cytoskeleton, which has been examined by the use of fluorescent-labelled antibody techniques (Derksen, Pierson & Traas, 1985) and freeze-substitution (Lancelle, Cresti & Hepler, 1987). Pollen tube microtubules are arranged in axial or S-helical arrays (Derksen *et al.*, 1985; Pierson, Derksen & Traas, 1986) and may be interspersed with and cross-linked to microfilaments (Lancelle *et al.*, 1987) forming a structurally integrated cytoskeletal complex with the plasma membrane (Tiwari & Polito, 1988). Numerous bundles of microfilaments are found in the tube cytoplasm, particularly a three-dimensional network of short microfilaments in the tip region (Tiwari & Polito, 1988). Their functions involve wall organization, the movement of the vegetative nucleus and the generative cell or sperm cells, vesicle movement, and movement of cell organelles during cytoplasmic streaming (Heslop-Harrison *et al.*, 1988). Myosin has recently been detected by immunofluorescence techniques on individual organelles in *Alopecurus pratensis* and *Secale cereale* and on the surfaces of vegetative nuclei and generative cells in *Hyacinthus orientalis* and *Helleborus foetidus* (Heslop-Harrison & Heslop-Harrison, 1989). These data indicate that the movement of cytoplasmic components in the pollen tube are driven by the interaction of myosin and actin fibrils at zones of contact (Heslop-Harrison & Heslop-Harrison, 1989). Contacts between the vegetative nucleus and the generative cell or sperm cells (Dumas *et al.*, 1988) may also increase the efficiency and speed of movement.

Pollen tube growth in the style is believed to be heterotrophic and follows the path of the available reserves through the transmitting tissue. Starch digestion is triggered by pollination and only occurs in compatible matings in *Petunia* (Herrero & Dickinson, 1979).

The time taken for the pollen tubes to reach the ovary varies greatly between species; this variation presumably indicates differences in style length, the form of the stylar transmitting tissue and ambient temperature. Heslop-Harrison (1987) has also suggested that the speed of wall synthesis is an important factor. Pollen tube growth rates are variable from 3 μm s^{-1} in maize and 2 μm s^{-1} in rye (Heslop-Harrison 1979b) to 0.3 – 1.4 μm s^{-1} in *Lilium longiflorum* (Van der Woude & Morre, 1967), *Talinum mengesii* (DuBay, 1981) and *Gasteria verrucosa*

(Willemse & Franssen-Verheijen, 1988) to $0.07 - 0.14$ μm s^{-1} in *Crocus* spp. (Heslop-Harrison, 1977) and *Rhododendron nuttallii* (Palser *et al.*, 1989). In all cases, however, the process appears to be dependent on the prevailing environmental conditions (Palser *et al.*, 1989).

A number of publications suggest that pollen tube growth is biphasic (Dickinson & Lawson, 1975) or triphasic (Willemse & Franssen-Verheijen, 1988). Initial growth, particularly penetration of the stigma and growth in the neck of the style, is slow, but rapidly increases in growth rate to a maximum from this point. Recently, Willemse & Franssen-Verheijen (1988) have shown that initial growth in *Gasteria verrucosa* is slow (0.3 μm s^{-1}). Growth through the style is more rapid (1.4 μm s^{-1}); growth in the ovary returns to a rate similar to the initial growth rate (0.3 μm s^{-1}). Such changes in growth rate must reflect either limitations in nutritional resource or rapid metabolic change induced by the changing environments.

The time taken for pollen tube growth from stigma to ovary varies from 30 to 60 min in *Crepis capillaris* (Gerassimova, 1933; Kuroiwa, 1989), *Triticum aestivum* (Bennett *et al.*, 1973) and *Hordeum vulgare* (Engell, 1988) to 4–5 h in *Gasteria verrucosa* (Willemse & Franssen-Verheijen, 1988), 16 h in *Salpiglossis sinuata* (Hepher & Boulter, 1987), 60 h in *Juglans regia* (Luza *et al.*, 1987), 8–9 d in *Rhododendron nuttallii* (Palser *et al.*, 1989) and 12 d in *Prunus persica* (Herrero & Arbeloa, 1989). In only the last species is the pollen tube seen to stop growth while the obturator becomes receptive to pollen tube passage (Herrero & Arbeloa, 1989).

Pollen tube guidance

The growth of the pollen tube in stigma, style and ovary requires nutritional support; the structures of each organ or tissue may serve to orientate the pollen tube (Hepher & Boulter, 1987). Webb & Williams (1988) suggest that the guidance system for pollen tube growth through the pistil is likely to combine at least three factors. Firstly, pollen tubes probably follow paths of least physical resistance and tend to grow along surfaces because of their thigmotropic characteristic (Hirouchi & Suda, 1975). Gasser *et al.* (1989) isolated several pistil-specific cDNAs, one of which is expressed at anthesis but only in tissues subtending the stigma and in the outer layers of the transmitting tissue. They hypothesize that the expressed gene product is involved in preventing pollen tubes from leaving the transmitting tissue during pollen tube growth. Secondly, the stylar transmitting tract is assumed to provide the appropriate physical medium, nutritional factors (Labarca & Loewus, 1972, 1973), and recognition factors. Thirdly, chemotropism may impose directional tube growth (Miki-Hiroshige, 1964; Welk, Millington & Rosen, 1965; Mascarenhas, 1975; Picton & Steer, 1983; Mulcahy & Mulcahy, 1986) although evidence for this is not convincing (Heslop-Harrison, 1986) except, perhaps, in the ovary (Webb & Williams, 1988). Calcium has been identified as acting as a possible chemotropic agent (Steer & Steer, 1989). In experiments designed to test the effects of pollination on stylar changes, Mulcahy & Mulcahy (1988) pollinated *Nicotiana alata* flowers in mid-style and observed pollen tube growth. In unpollinated stigmas, pollen tubes grew both towards the stigmatic end and towards the ovary, suggesting no chemotropic gradient, whereas in pollinations performed after a stigmatic one, pollen tubes grew predominantly towards the ovary, suggesting that a gradient had been set up.

Feedback mechanisms

In *Kleinhovia hospita*, a unique mechanism of stylar plugging by the early-formed zygotes has been reported (Shaanker & Ganeshaiah, 1989). The authors consider that stylar plugging is mediated by IAA released from fertilized ovules, which diffuses back into the style and blocks pollen germination and tube growth.

   Primack (1985) has shown accelerated senescence of flowers of greenhouse-grown plants following pollination. More recently, Richardson & Stephenson (1989) have shown, by using hand and field pollinations in the protandrous species *Campanula rapunculoides*, that the source of pollen (self or cross) delays the senescence of the flower, resulting in an extension of the pistillate phase. This effect was also noted where the number of fertilized ovules was low in any one flower.

Pollen tube growth in the ovary and entry to the micropyle

Entry into the ovule from the transmitting zone is followed by tube growth over the ovarian transmitting zone (placental epithelium), which is intimately associated with the micropyle of the ovule and facilitates pollen tube contact. The transmitting zone may (Cass *et al.*, 1986; Clifford & Owens, 1990) or may not (Olson, 1988) be associated with placental exudate. The placentae of *Papaver nudicaule* produce an exudate through which pollen tubes must grow on their way to the ovular micropyles (Cass *et al.*, 1986); the exudate may be directly involved in the recognition of the compatible tube. The placentae of *Solanum tuberosum* are stimulated by successful pollination; placental cell proliferation and enlargement ensue (Olson, 1988).

Herrero, Arbeloa & Gascon (1988) reported that the period from arrival of pollen tubes at the base of the style to fertilization was variable in the peach (*Prunus persica*) and that this period could extend beyond the time taken for pollen-tube growth down the style. The principal cause of the delay appears to be the obturator, a placental protuberance connecting the style with the micropyle of the ovule, which at this time appears unreceptive to pollen tubes. Starch digeston takes place in the obturator and is concomitant with secretion of proteins and carbohydrates; however, this development is not stimulated by pollination but appears to be a maturation process, since it occurs in both pollinated and unpollinated pistils (Arbeloa & Herrero, 1987). Hence pollen tubes must wait for its receptivity. Once pollen tubes have passed the obturator, callose begins to be deposited in this structure (Arbeloa & Herrero, 1987). This latter paper is the only demonstration of such a phenomenon, other workers observing pollen tube growth on the obturator but not observing such changes (Tilton & Horner, 1980; Tilton *et al.*, 1984; Hill & Lord, 1987).

Once the pollen tubes have traversed the obturator, there are further periods before micropylar entry and fertilization (Herrero *et al.*, 1988). A chemotrophic effect may be necessary for penetration of the ovule (Lord & Kohorn, 1986).

Fertilization

Pollen tube entry into the embryo sac

Generally, pollen tube entry into the micropyle is followed by intercellular growth through the nucellus to the embryo sac. Various modifications and elaborations of the micropylar area exist, such as the extension of the inner integument in *Crocus* to form a micropylar canal (Rudall, Owens & Kenton, 1984). The most bizarre cases, however, are where the micropylar ends of the synergid cells may be extended into the micropyle (and even up the style sometimes to the stigma in Viscaceae; Maheshwari, 1950) or modified into hooks. The embryo sac cell first encountered is generally one of the synergids. The cell wall at the micropylar end is often elaborated into the filiform apparatus, a labyrinthine wall modification of structure similar to that in transfer cells. Such a structure has led some authors to consider that these cells are necessarily involved in secretion and absorption (van Went, 1970; Chao, 1971; Wilms, 1981). A filiform apparatus is absent in *Crepis capillaris* (Kuroiwa, 1989).

Recently, from a study of 35 000 ovules by using ovule-clearing techniques in *Rhododendron nuttallii*, it has become apparent that the ovule and its enclosed gametophyte are far from mature during the period of anthesis and pollination (Palser *et al.*, 1989). The earliest stage observed at anthesis was a linear tetrad; the majority of megagametophytes were at a stage earlier than four nucleate. In fact, no megagametophyte was mature at pollination. Some of the variation may be due to ovule position on the placenta (Palser *et al.*, 1989).

The order of pollen tube entries into ovules and subsequent fertilization may be related to variations in the rate of gametophyte maturation in different ovules as well as variations in the rate of pollen tube growth. Fertilized ovules constituted less than 5% of the total ovules until 12 d after pollination (Palser *et al.*, 1989).

A number of investigations have indicated that pollination stimulates the degeneration of one of the two synergids (Table 2) and that pollen tube entry and sperm discharge in the embryo sac must occur via the degenerate synergid. The full story is yet to be revealed for variations in the structure and composition of the embryo sac are well known. The embryo sac may have one or no synergids (*Plumbago capensis*, Cass, 1972; *P. zeylanica*, Russell, 1982), and ephemeral through to multiple antipodal cells (e.g. Gramineae).

Penetration by more than one pollen tube is occasionally observed (Palser *et al.*, 1989). This has also been observed where wide crosses have been attempted between wheat and maize (Laurie & Bennett, 1988).

Gamete delivery

The time taken between pollination and fertilization varies, not surprisingly, between families, genera and species. The shortest time recorded is 15–45 min in *Taraxacum koksaghyz* (Poddubnaya-Arnoldii & Dianowa, 1934) and the longest 6–7 months in *Corylus* (Lagerstedt, 1979; Thompson, 1979) and orchids (Clifford & Owens, 1988), and even 12–14 months in species of *Quercus* (Bagda, 1948).

The fertilization process has been thoroughly reviewed by van Went & Willemse (1984). It is briefly outlined here with the addition of new data where appropriate. Ultrastructural studies show that pollen tubes grow into and through the filiform apparatus. In *Rhododendron nuttallii* the pollen tube enters the ovary through the micropyle and penetrates one of the synergids, generally from the side and away from the densest region of the filiform apparatus. No change in synergids is discernable before penetration of the tube, but changes occur subsequently (Palser *et al.*, 1989). There is a decrease in cell volume, collapse of vacuoles, disintegra-

Table 2. *The effect of pollination, pollen tube growth and fertilization on the synergids*

Species	Synergid degeneration			Reference
	pollination not required	pollination required	pollen tube entry required	
Capsella bursa-pastoris	–	–	+	Schultz & Jensen, 1968
Crepis capillaris	–	+	–	Kuroiwa, 1989
Epidendrum scutella	–	+	–	Coccuci & Jensen, 1969
Gossypium hirsutum	–	+	–	Jensen, 1965; Jensen & Fisher, 1968
Helianthus annuus	–	–	+	Newcomb, 1973
Hordeum vulgare	+	+	–	Engell, 1988; Cass & Jensen, 1970; Mogensen, 1988
Linum usitatissimum	–	+	–	Vazart, 1971
Lycopersicon esculentum	–	+	–	Nettancourt et al., 1973
Nicotiana tabacum	–	+	–	Mogensen & Suthar, 1979
Paspalum orbiculare	–	+	–	Chao, 1971
Petunia hybrida	–	–	+	van Went, 1970
Quercus gambelii	–	–	+	Mogensen, 1972
Spinacia oleracea	–	+	–	Wilms, 1981
Stipa elmeri	–	+	–	Maze & Lin, 1975
Torenia fournieri	–	–	+	van der Pluym, 1964
Zea mays	–	+	–	Diboll, 1968

tion of the plasma membrane, and disorganization of the cell organelles. Pollen-tube growth now ceases and the tube contents are discharged either in a controlled fashion through a pore (Jensen & Fisher, 1968) or by the tube simply bursting. In *Petunia hybrida*, discharge may also rupture the synergid (van Went, 1970).

The delivery of gametes in *Plumbago zeylanica* (Russell & Cass, 1981; Russell, 1982) and *Plumbagella micrantha* (Russell & Cass, 1983) is unusual, as the embryo sac has no synergids. The pollen tube enters the embryo sac through a filiform apparatus at the base of the egg cell, at the junction between the egg cell and the central cell. Sperms are released from a terminal pore in the pollen tube and evidently transmitted, one to the egg and the other to the central cell.

In the majority of cases, the male gametes are transferred simply as a nucleus and are entirely lacking cytoplasm (Hagemann & Schroder, 1989). In wheat, barley and *Triticale*, generative and sperm cells in the pollen grain contain plastids (Schroder, 1983; Hagemann & Schroder, 1985) but there is no evidence of biparental plastid inheritance (Hagemann & Schroder, 1989). This result indicates that sperm cytoplasm is cleaned from the nucleus and not transmitted to the zygote, a situation observed for barley by Mogensen (1988). Similarly, in *Spinacia*, where plastids are not incorporated into sperm cells, sperm cytoplasm remains outside egg and central cells (Wilms, 1981). There is no evidence that plastids are transmitted to the egg cell and then destroyed (Hagemann & Schroder, 1989).

There is a minority of species (*Oenothera erythrosepala* (Meyer & Stubbe, 1974), *O. hookeri* (Diers, 1963), *Pelargonium zonale* (Lombardo & Gerola, 1968; Khera, 1975), *Plumbago zeylanica* (Russell & Cass, 1983), and *Rhododendron* species (Knox & Williams, 1986)) where transfer does involve nuclei and plastids and presumably other cytoplasmic components. Genetic analysis strongly supports biparental inheritance (Kirk & Tilney-Bassett, 1978). In *Plumbago zeylanica* (Russell & Cass, 1983), there are indications that one sperm cell, carrying all the plastids, is preadapted to fusion with the egg, while the other sperm, lacking plastids, fuses with the polar nuclei. Research is currently aimed at surface membrane receptors on sperms, which may play a role in directing sperm to target cell and nucleus (Matthys-Rochon & Dumas, 1988). There is no evidence to indicate the fate of other cytoplasmic components.

In *Triticum aestivum*, Bennett *et al.* (1973) report that the pollen tube discharges the sperm nuclei either into the egg cell or close to it, and the second sperm nucleus migrates to the central cell and polar nuclei. Delivery and migration to the egg nucleus is therefore first.

The common border between the synergids and the egg and central cells comprises the plasma membranes of each cell with intermittent wall interruptions along their length (Kuroiwa, 1989). This situation undoubtedly facilitates delivery of sperm cells.

Nuclear fusion

Very little is known about the fusion of the nuclei of egg and sperm and polar nuclei and sperm. Fusion of the polar nuclei in the central cell may occur before fusion with the sperm nucleus or fusion may be simultaneous. At fertilization the polar nuclei are still separate in *Rhododendron nuttallii* (Palser *et al.*, 1989). The sperm nucleus becomes increasingly diffuse before it fuses with the egg nucleus in *Triticum aestivum* (Bennett *et al.*, 1973). No sperm nucleus could be recognized between 18 and 20 h after pollination. In the central cell, the polar nuclei and sperm nucleus remained distinct at 2–3 h after pollination and still had separate nuclear membranes. At prophase of the first division, the sperm chromosomes remained distinct but increasingly merged with the polar nuclei as contraction progressed. Division was apparent at 5 h (Bennett *et al.*, 1973). The observations of Bennett *et al.* (1973) differed from those of Morrison (1955), who reported that chromosomes were not visible as separate groups at primary division of the endosperm nucleus.

In *Triticum aestivum*, the first division of the zygote is seen at about 22 h from pollination; by 24 h after pollination, the endosperm consists of about 16 nuclei (Bennett *et al.*, 1973). In *Hordeum vulgare* (Engell, 1988), sperm and egg nuclei were fused by 45 min after pollination, the first zygotic division being timed similarly to wheat at 22–24 h after pollination. Following the development of the single primary endosperm nucleus, a transitory callose wall is formed around the zygote in *Rhododendron* and *Ledum* species (Williams *et al.*, 1984; Palser *et al.*, 1989).

Conspectus

The pistil undergoes preprogrammed changes during development, including many changes post-anthesis, which affect pollen adherence, pollen germination and pollen-tube growth. Pollen touch-down itself may set in train a sequence of post-pollination responses in stigma, style, ovary, micropyle and embryo sac, which may or may not lead to gamete delivery and fertilization. Several post-pollination responses have now been identified although their distribution among species is not fully known. Their significance is being evaluated for different breeding systems in different plant groups but they are of particular interest to those manipulating breeding systems and to those involved in wide crossing.

Current work on the molecular biology of both gametophytic and sporophytic self-incompatibility is providing new insights into the means by which the process of self-fertilization is controlled (Lawrence, 1989) but our understanding of self-incompatibility systems is still far from complete. In gametophytic self-incompatibility, for example, the occurrence of late-acting self-incompatibility (Seavey & Bawa, 1986; Owens, 1985) together with multigene control (Brandham & Owens, 1978) may be more frequent than acknowledged at present. It is clear that a number of controls, additional to self-incompatibility, exist; these controls block the success of crosses between species and genera. Clearly there is an individual genotypic influence (O'Donoughue & Bennett, 1988; Laurie & Bennett, 1988) as well as individual genes (*Kr* genes in wheat) (Lange & Wojceichowska, 1976; Jalani and Moss, 1980; Snape *et al.*, 1979; Sitch, 1984; Thomas, Kaltsikes & Anderson, 1981; Laurie & Bennett 1987) controlling these effects.

References

Arbeloa, A. & Herrero, M. (1987). The significance of the obturator in the control of pollen tube entry into the ovary in peach (*Prunus persica*). *Annals of Botany* **50**, 681–5.

Arditti, J. (1979). Aspects of the physiology of orchids. *Advances in Botanical Research* **7**, 421–655.

Bagda, H. (1948). Morphologische und biologische Untersuchungen über Valonea Eichen (*Quercus macropelis* Ky.) in Haci-Kadin-Tal bei Ankara. *Communications of the Faculty of Science, University of Ankara* **1**, 89–125. (Cited in Maheshwari, 1950.)

Bennett, M.D., Rao, M.K., Smith, J.B. & Bayliss, M.W. (1973). Cell development in the anther, the ovule and the young seed of *Triticum aestivum* L. var. Chinese Spring. *Philosophical Transactions of the Royal Society of London* **266**, 39–81.

Bertin, R.I. (1986). Consequences of mixed pollinations in *Campsis radicans*. *Oecologia* **70**, 1–5.

Brandham, P.E. & Owens, S.J. (1978). The genetic control of self-incompatibility in the genus *Gasteria* (Liliaceae). *Heredity* **40**, 165–9.

Cass, D.D. (1972). Occurrence and the development of a filiform apparatus in the egg of *Plumbago capensis*. *American Journal of Botany* **59**, 279–83.

Cass, D.D., Ashley, C., Folsum, M.W. & Fossum, A.M. (1986). Further experiments with *in vitro* pollination of *Papaver nudicaule*. In *Pollination '86* (ed. E.G. Williams & R.B. Knox), pp. 155–61. School of Botany, University of Melbourne.

Cass, D.D. & Jensen, W.A. (1970). Fertilization in barley. *American Journal of Botany* **57**, 62–70.

Chao, Ch.Y. (1971). A periodic acid-Schiff's substance related to the directional growth of pollen tube into embryo sac in *Paspalum* ovules. *American Journal of Botany* **58**, 649–54.

Clarke, A.E., Anderson, M.A., Atkinson, A., Bacic, A., Ebert, P.R., Jahnen, W., Lush, W.M., Mau, S.-L. & Woodward, J.R. (1989). Recent developments in the molecular genetics and biology of self incompatibility. *Plant Molecular Biology* **13**, 267–71.

Clifford, S.C. & Owens, S.J. (1988). Post-pollination phenomena and embryo development in the Oncidiinae (Orchidaceae). In *Sexual reproduction in higher plants* (ed. M. Cresti, P. Gori & E. Pacini), pp. 407–12. Berlin: Springer-Verlag.

Clifford, S.C. & Owens, S.J. (1990). The stigma, style and ovarian transmitting tract in the Oncidiinae (Orchidaceae): morphology, developmental anatomy, and histochemistry. *Botanical Gazette* **151**, 440–51.

Coccuci, A. & Jensen, W.A. (1969). Orchid embryology: mega-gametophyte of *Epidendrum scutella* following fertilization. *American Journal of Botany* **56**, 629–40.

Cornish, E.C., Anderson, M.A. & Clarke, A.E. (1988). Molecular aspects of fertilization in flowering plants. *Annual Review of Cell Biology* **4**, 209–28.

Cresti, M.E., Pacini, E., Ciampolini, F. & Sarfatti, G. (1977). Germination and pollen tube development *in vitro* of *Lycopersicon peruvianum* pollen: ultrastructural features. *Planta* **136**, 239–47.

Derksen, J., Pierson, E.S. & Traas, J.A. (1985). Microtubules in vegetative and generative cells of pollen tubes. *European Journal of Cell Biology* **38**, 142–8.

Diboll, A.G. (1968). Fine structure development of the mega-gametophyte of *Zea mays* following fertilization. *American Journal of Botany* **55**, 787–806.

Dickinson, H.G. & Elleman, C.J. (1985). Structural changes in the pollen grain of *Brassica oleracea* during dehydration in the anther and development on the stigma as revealed by anhydrous fixation techniques. *Micron and Microscopica Acta* **16**, 255–70.

Dickinson, H.G. & Lawson, J. (1975). Pollen tube growth in the stigma of *Oenothera organensis* following compatible and incompatible intraspecific pollinations. *Proceedings of the Royal Society of London* **B188**, 327–44.

Dickinson, H.G., Moriarty, J. & Lawson, J. (1982). Pollen–pistil interaction in *Lilium longiflorum*: the role of the pistil in controlling pollen tube growth following cross and self pollinations. *Proceedings of the Royal Society of London* **B215**, 45–62.

Diers, L. (1963). Electronenmikroskopische Beobachtungen an der generativen Zelle von *Oenothera hookeri* Torr. et Gray. *Zeitschrift für Naturforschung* **18**, 562–6.

DuBay, D.T. (1981). *Interspecific differences in the effects of sulphur dioxide on angiosperm sexual reproduction.* Ph.D. dissertation, Emory University, Atlanta, Georgia.

Dumas, C., Bowman, R.B., Gaude, T., Guilly, C.M., Heizmann, Ph., Roeckel, P. & Rougier, M. (1988). Stigma and stigma secretion re-examined. *Phyton (Austria)* **28**, 193–200.

Elleman, C.J., Willson, C.E., Sarker, R.H. & Dickinson, H.G. (1988). Interaction between the pollen tube and the stigmatic cell wall following pollination in *Brassica oleracea. New Phytologist* **109**, 111–17.

Engell, K. (1988). Embryology of barley. II. Synergids and egg cell, zygote and embryo development. In *Sexual Reproduction in Higher Plants* (ed. M. Cresti, P. Gori & E. Pacini), pp. 383–8. Berlin: Springer-Verlag.

Ferrari, T.E., Best, V., More, T.A., Comstock, P., Muhammad, A. & Wallace, D.H. (1985). Intercellular adhesions in the pollen–stigma system: Pollen capture, grain binding, and tube attachments. *American Journal of Botany* **72**, 1466–74.

Fougere-Rifot, M. (1987). Micro- and mega-sporogenesis, fertilization, embryogenesis, and endosperm formation of Angiosperms. *Bulletin de la Société Botanique de France* **134**, 113–68.

Gasser, C.S., Budelier, K.A., Smith, A.G., Shah, D.M. & Fraley, R.T. (1989). Isolation of tissue-specific cDNAs from tomato pistils. *Plant Cell* **1**, 15–24.

Gerassimova, H. (1933). Fertilization in *Crepis capillaris. Cellule* **42**, 103–48.

Gibbs, P.E. (1986). Do homomorphic and heteromorphic self incompatibility systems have the same sporophytic mechanism? *Plant Systematics and Evolution* **154**, 285–323.

Hagemann, R. & Schroder, M.-B. (1985). New results about the presence of plastids in generative and sperm cells of Gramineae. In *Sexual Reproduction in Seed Plants, Ferns, and Mosses* (ed. M.T.M. Willemse & J.L. van Went), pp. 53–5. Wageningen: PUDOC.

Hagemann, R. & Schroder, M.-B. (1989). The cytological basis of the plastid inheritance in angiosperms. *Protoplasma* **152**, 57–64.

Hepher, A. & Boulter, M.E. (1987). Pollen tube growth and fertilization efficiency in *Salpiglossis sinuata*: Implications for the involvement of chemotrophic factors. *Annals of Botany* **60**, 595–601.

Herrero, M. & Arbeloa, A. (1989). Influence of the pistil on pollen tube kinetics in peach (*Prunus persica*). *American Journal of Botany* **76**, 1441–7.

Herrero, M., Arbeloa, A. & Gascon, M. (1988). Pollen pistil interaction in the ovary in fruit trees. In *Sexual Reproduction in Higher Plants* (ed. M. Cresti, P. Gori & E. Pacini), pp. 297–302. Berlin: Springer-Verlag.

Herrero, M. & Dickinson, H.G. (1979). Pollen–pistil incompatibility in *Petunia hybrida*: changes in the pistil following compatible and incompatible intraspecific crosses. *Journal of Cell Science* **36**, 1–18.

Herscovitch, J.C. & Martin, A.R.H. (1989). Pollen–pistil interactions in *Grevillea banksii*. The pollen grain, stigma, transmitting tissue, and in vitro pollinations. *Grana* **28**, 69–84.

Heslop-Harrison, J. (1979*a*). Aspects of the structure, cytochemistry and germination of the pollen of rye (*Secale cereale*). *Annals of Botany* **44** (suppl. 1), 1–47.

Heslop-Harrison, J. (1979*b*). Pollen–stigma interaction in grasses: a brief review. *New Zealand Journal of Botany* **17**, 537–46.

Heslop-Harrison, J. (1982). Pollen–stigma interaction in the Leguminosae: constituents of the stylar fluid and stigma secretion of *Trifolium pratense* L. *Annals of Botany* **49**, 729–35.

Heslop-Harrison, J. (1983). Self incompatibility: phenomenology and physiology. *Proceedings of the Royal Society* **B218**, 317–95.

Heslop-Harrison, J. (1986). Pollen tube chemotropism: fact or delusion? In *Biology of Reproduction and Cell Motility in Plants and Animals* (ed. M. Cresti & R. Dallai), pp. 169–74. Siena, Italy: University of Siena Press.

Heslop-Harrison, J. (1987). Pollen germination and pollen-tube growth. *International Review of Cytology* **107**, 1–78.

Heslop-Harrison, J. & Heslop-Harrison, Y. (1975). Fine structure of the stigmatic papilla of *Crocus*. *Micron* **6**, 45–52.

Heslop-Harrison, J. & Heslop-Harrison, Y. (1983). Pollen–stigma interaction in the Leguminosae: the organization of the stigma in *Trifolium pratense* L. *Annals of Botany* **51**, 571–83.

Heslop-Harrison, J. & Heslop-Harrison, Y. (1989). Myosin associated with the surfaces of organelles, vegetative nuclei and generative cells in angiosperm pollen grains and tubes. *Journal of Cell Science* **94**, 319–25.

Heslop-Harrison, J., Heslop-Harrison, Y., Cresti, M., Tiezzi, A. & Moscatelli, A. (1988). Cytoskeletal elements, cell shaping and movement in the angiosperm pollen tube. *Journal of Cell Science* **91**, 49–60.

Heslop-Harrison, Y. (1977). The pollen–stigma interaction: pollen tube penetration in *Crocus*. *Annals of Botany* **41**, 913–22.

Heslop-Harrison, Y. (1981). Stigma characteristics and angiosperm taxonomy. *Nordic Journal of Botany* **1**, 401–20.

Heslop-Harrison, Y. & Shivanna, K.R. (1977). The receptive surface of the angiosperm stigma. *Annals of Botany* **41**, 1233–58.

Hill, J.P. & Lord, E.M. (1987). Dynamics of pollen tube growth in the wild radish *Raphanus raphanistrum* (Brassicaceae). II. Morphology, cytochemistry and ultrastructure of transmitting tissues, and the path of pollen tube growth. *American Journal of Botany* **74**, 987–8.

Hirouchi, T. & Suda, S. (1975). Thigmotropism in the growth of pollen tubes of *Lilium longiflorum*. *Plant and Cell Physiology* **16**, 337–81.

Hoekstra, F.A. (1979). Mitochondrial development and activity of binucleate and trinucleate pollen during germination *in vitro*. *Planta* **145**, 25–36.

Hoekstra, F.A. & Bruinsma, J. (1979). Protein synthesis of binucleate and trinucleate pollen and its relationship to tube emergence and growth. *Planta* **146**, 559–66.

Jalani, B.S. & Moss, J.P. (1980). The site of action of the crossability genes (Kr1, Kr2) between *Triticum* and *Secale*. I. Pollen germination, pollen tube growth, and number of pollen tubes. *Euphytica* **29**, 571–9.

Jensen, W.A. (1965). The ultrastructure and histochemistry of the synergids of cotton. *American Journal of Botany* **52**, 238–56.

Jensen, W.A. & Fisher, J.B. (1968). Cotton embryogenesis: The entrance and discharge of the pollen tube into the embryo sac. *Planta* **78**, 158–83.

Kenrick, J. & Knox, R.B. (1981). Post-pollination exudate from stigmas of *Acacia* (Mimosaceae). *Annals of Botany* **48**, 103–6.

Khera, P.K. (1975). *Plastid development in zonal pelargoniums*. Ph.D. thesis, University of Wales, Swansea.

Kirk, J.T.O. & Tilney-Bassett, R.A.E. (1978). *The Plastids. Their Chemistry, Structure, Growth and Inheritance*. Amsterdam: Elsevier/ North Holland.

Knox, R.B. (1984). Pollen–pistil interactions. In *Cellular Interactions* (*Encyclopedia of Plant Physiology*, new series, vol. 17) (ed. H.F. Liskens & J. Heslop-Harrison), pp. 508–608. Berlin: Springer-Verlag.

Knox, R.B. & Williams, E.G. (1986). Pollen, pistil and reproductive function in crop plants. *Plant Breeding Review* **4**, 9–79.

Kreitner, G.L. & Sorenson, E.L. (1984). Stigma development and the stigmatic cuticle of alfalfa, *Medicago sativa* L. *Botanical Gazette* **145**, 436–43.

Kuroiwa, H. (1989). Ultrastructural examination of embryogenesis in *Crepis capillaris* (L.) Wallr. 1. The synergid before and after pollination. *Botanical Magazine of Tokyo* **102**, 9–24.

Labarca, C. & Loewus, F. (1972). The nutritional role of pistil exudate in pollen tube wall formation in *Lilium longiflorum*. I. Utilization of injected stigmatic exudate. *Plant Physiology* **50**, 7–14.

Labarca, C. & Loewus, F. (1973). The nutritional role of pistil exudate in pollen tube wall formation in *Lilium longiflorum*. II. Production and utilization of exudate from stigma and stylar canal. *Plant Physiology* **52**, 87–92.

Lagerstedt, H.B. (1979). Filberts. In *Nut tree culture in North America* (ed. R.A. Jaynes), pp. 128–47. Harriden, Connecticut: Northern Nut Tree Growers Association.

Lancelle, S.A., Cresti, M. & Hepler, P.K. (1987). Ultrastructure of the cytoskeleton in freeze-substituted pollen tubes of *Nicotiana alata*. *Protoplasma* **140**, 141–50.

Lange, W. & Wojciechowska, B. (1976). The crossing of common wheat (*Triticum aestivum* L.) with cultivated rye (*Secale cereale* L.). I. Crossability, pollen germination and pollen tube growth. *Euphytica* **25**, 609–20.

Laurie, D.A. & Bennett, M.D. (1987). The effect of the crossability loci Kr1 and Kr2 on fertilization frequency in hexaploid wheat × maize crosses. *Theoretical and Applied Genetics* **73**, 403–9.

Laurie, D.A. & Bennett, M.D. (1988). Chromosome behaviour in wheat × maize, wheat × sorghum and barley × maize crosses. In *The Kew Chromosome Conference*, vol. 3 (ed. P.E. Brandham), pp. 167–77. London: HMSO.

Lawrence, M.J. (1989). Molecular genetics of self incompatibility. *Genome* **31**, 476–7.

Lombardo, G. & Gerola, F.M. (1968). Cytoplasmic inheritance and ultrastructure of the male generative cell in higher plants. *Planta* **82**, 105–10.

Lord, E.M. & Heslop-Harrison, Y. (1984). Pollen–stigma interaction in the Leguminosae: Stigma organization and breeding system in *Vicia faba* L. *Annals of Botany* **54**, 827–36.

Lord, E.M. & Kohorn, L.V. (1986). Gynoecial development, pollination, and the path of the pollen tube in the tepary bean, *Phaseolus acutifolius*. *American Journal of Botany* **73**, 70–8.

Lovell, P.J., Lovell, P.H. & Nichols, R. (1987). The control of flower senescence in Petunia (*Petunia hybrida*). *Annals of Botany* **60**, 49–59.

Luza, J.G., Polito, V.S. & Weinbaum, S.A. (1987). Staminate bloom date and temperature responses of pollen germination and tube growth in two walnut (*Juglans*) species. *American Journal of Botany* **74**, 1898–903.

McClure, B.A., Haring, V., Ebert, P.R., Anderson, M.A., Simpson, R.J., Sakiyama, F. & Clarke, A.E. (1989). Style self-incompatibility products of *Nicotiana alata* are ribonucleases. *Nature* **342**, 955–7.

Maheshwari, P. (1950). *An Introduction to the Embryology of the Angiosperms*. New York: McGraw-Hill.

Marshall, D.L. & Ellstrand, N.C. (1986). Sexual selection in *Raphanus sativus*: experimental data on non-random fertilization, maternal choice, and consequences of multiple paternity. *American Naturalist* **127**, 446–61.

Mascarenhas, J.P. (1975). The biochemistry of Angiosperm pollen development. *Botanical Review* **41**, 259–314.

Mascarenhas, J.P. (1989). The male gametophyte of flowering plants. *Plant Cell* **1**, 657–64.

Matthys-Rochon, E. & Dumas, C. (1988). The male germ unit: retrospect and prospects. In *Plant Sperm Cells as Tools for Biotechnology* (ed. H.J. Wilms & C.J. Keijzer), pp. 51–60. Wageningen: Pudoc.

Maze, J. & Lin, S.C. (1975). A study of the mature megagametophyte of *Stipa elmeri*. *Canadian Journal of Botany* **53**, 2958–77.

Meyer, B. & Stubbe, W.W. (1974). Das Zahlenverhaltnis von mutterlichen und vaterlichen Plastiden in den Zygoten von *Oenothera erythrosepala* Borbas (syn. *Oe. lamarkiana*). *Berichte der Deutschen Botanischen Gesellschaft* **87**, 29–38.

Miki-Hiroshige, H. (1964). Tropism of pollen tubes to the pistils. In *Pollen Physiology and Fertilization* (ed. H.F. Linskens), pp. 152–8. Amsterdam: North-Holland.

Mogensen, H.L. (1972). Fine structure and composition of the egg apparatus before and after fertilization in *Quercus gambellii*: the functional ovule. *American Journal of Botany* **59**, 931–41.

Mogensen, H.L. (1988). Exclusion of male mitochondria and plastids during syngamy as a basis for maternal inheritance. *Proceedings of the National Academy of Sciences, USA* **85**, 2594–7.

Mogensen, H.L. & Suthar, H.K. (1979). Ultrastructure of the egg apparatus of *Nicotiana tabacum* (Solanaceae) before and after fertilization. *Botanical Gazette* **140**, 169–79.

Morrison, J.W. (1955). Fertilization and post fertilization development in wheat. *Canadian Journal of Botany* **33**, 168–76.

Mulcahy, G.B. & Mulcahy, D.L. (1986). More evidence on the preponderant influence of the pistil on pollen tube growth. In *Biology of Reproduction and Cell Motility in Plants and Animals* (ed. M. Cresti & R. Dallai), pp. 139–44. Siena, Italy: University of Siena Press.

Mulcahy, G.B. & Mulcahy, D.L. (1988). Induced polarity as an index of pollination-triggered stylar activity. In *Sexual Reproduction in Higher Plants* (ed. M. Cresti, P. Gori & E. Pacini), pp. 327–32. Berlin: Springer-Verlag.

Murdy, W.H. & Carter, M.E.B. (1987). Regulation of the timing of pollen germination by the pistil in *Talinum mengesii* (Portulacaceae). *American Journal of Botany* **74**, 1888–92.

Nettancourt, D. de, Devreux, H., Laneri, N., Pacini, E., Cresti, M. & Sarfatti, G. (1973). Ultrastructural aspects of bilateral interspecific incompatibility between *Lycopersicum peruvianum* and *Lycopersicum esculentum*. *Caryologia* **25**, 207–17.

Newcomb, W. (1973). The development of the embryo sac of the sunflower *Helianthus annuus* after fertilization. *Canadian Journal of Botany* **51**, 879–90.

O'Donoughue, L.S. & Bennett, M.D. (1988). Wide hybridization between relatives of bread wheat and maize. In *Proceedings of 7th International Wheat Genetics Symposium*, Cambridge (ed. T.E. Miller & R.M.D. Koebner), pp. 397–402. Cambridge: Institute of Plant Science Research.

Olson, A.R. (1988). Postpollination placental development of a diploid *Solanum tuberosum*. *Canadian Journal of Botany* **66**, 1813–17.

Owens, S.J. (1985). Seed set in *Lotus berthelotii* Masferrer. *Annals of Botany* **55**, 811–14.

Owens, S.J. (1989). Stigma, style, pollen, and the pollen–stigma interaction in Caesalpinioideae. In *Advances in Legume Biology* (Monographs of the Missouri Botanical Gardens no. 29) (ed. C.J.

Sirton & J. Zarruchi), pp. 113–26. St Louis: Missouri Botanical Garden.

Palser, B.F., Rouse, J.L. & Williams, E.G. (1989). Coordinated time-tables for megagametophyte development and pollen tube growth in *Rhododendron nuttallii* from anthesis to early postfertilization. *American Journal of Botany* **76**(8), 1167–202.

Picton, J.M. & Steer, M.W. (1983). Evidence for the role of Ca2+ ions in tip extension in pollen tubes. *Protoplasma* **115**, 11–17.

Pierson, E.S., Derksen, J. & Traas, J.A. (1986). Organization of micro-filaments and microtubules in pollen tubes in vitro and in vivo in various angiosperms. *European Journal of Cell Biology* **41**, 14–18.

Pluym, J.E. van der (1964). An electron microscopic investigation of the filiform apparatus in the embryo sac of *Torenia fournieri*. In *Pollen Physiology and Fertilization* (ed. H.F. Linskens), pp. 6–16. Amsterdam: North-Holland.

Poddubnaya-Arnoldii, V.A. & Dianowa, V. (1934). Ein zytoembryolo-gische Untersuchengen einiger Arten der Gattung *Taraxacum*. *Planta* **23**, 16–46.

Primack, R.B. (1985). Longevity of individual flowers. *Annual Review of Ecology and Systematics* **16**, 15–37.

Richards, A.J. (1986). *Plant Breeding Systems*. London: George Allen & Unwin.

Richardson, T.E. & Stephenson, A.G. (1989). Pollen removal and pol-len deposition affect the duration of the staminate and pistillate phases in *Campanula rapunculoides*. *American Journal of Botany* **76**, 532–8.

Roberts, I.N., Stead, A.D., Ockenden, D.J. & Dickinson, H.G. (1980). Pollen stigma interactions in *Brassica oleracea*. *Theoretical and Applied Genetics* **58**, 241–6.

Rudall, P.J., Owens, S.J. & Kenton, A.Y. (1984). Embryology and breeding systems in *Crocus* (Iridaceae) – A study in causes of chromo-some variation. *Plant Systematics and Evolution* **148**, 119–34.

Russell, S.D. (1982). Fertilization in *Plumbago zeylanica*: Entry and discharge of the pollen tube into the embryo sac. *Canadian Journal of Botany* **60**, 2219–30.

Russell, S.D. & Cass, D.D. (1981). Ultrastructure of fertilization in *Plumbago zeylanica*. *Acta Societatis Botanicorum Poloniae* **50**, 185–9.

Russell, S.D. & Cass, D.D. (1983). Unequal distribution of plastids and mitochondria during sperm cell formation in *Plumbago zeylanica*. In *Pollen: Biology and Implications for Plant Breeding* (ed. D.L. Mulcahy & E. Ottaviano), pp. 135–40. New York: Elsevier.

Schemske, D.W. & Pautler, L.P. (1984). The effects of pollen composi-tion on fitness components in a neotropical herb. *Oecologia* **62**, 31–6.

Schroder, M.-B. (1983). The ultrastructure of sperm cells in *Triticale*. In: *Fertilization and Embryogenesis in Ovulated Plants* (ed. O. Erdelska), pp. 101–4. Bratislava: Veda.

Schultz, R. & Jensen, W.A. (1968). *Capsella* embryogenesis: the synergids before and after fertilization. *American Journal of Botany* **55**, 541–52.

Seavey, S.R. & Bawa, K.S. (1986). Late acting self incompatibility in angiosperms. *Botanical Review* **52**, 196–219.

Sedgely, M. & Smith, R.M. (1989). Pistil receptivity and pollen tube growth in relation to the breeding system of *Eucalyptus woodwardii* (*Symphyomyrtus*: Myrtaceae). *Annals of Botany* **64**, 21–31.

Shaanker, R.U. & Ganeshaiah, K.N. (1989). Stylar plugging by fertilized ovules in *Kleinhovia hospita* (Sterculiaceae) – a case of vaginal sealing in plants? *Evolutionary Trends in Plants* **3**, 59–64.

Sitch, L.A. (1984). *The production and utilization of wheat doubled haploids*. Ph.D. thesis, University of Cambridge.

Snape, J.W., Chapman, V., Moss, J., Blanchard, C.E. & Miller, T.E. (1979). The crossabilities of wheat varieties with *Hordeum bulbosum*. *Heredity* **42**, 291–8.

Steer, M.W. & Steer, J.M. (1989). Pollen tube tip growth. *New Phytologist* **111**, 323–58.

Thomas, J.B., Kaltsikes, P.J. & Anderson, R.G. (1981). Relation between wheat–rye crossability and seed set of common wheat after pollination with other species in the Hordeae. *Euphytica* **30**, 121–7.

Thompson, M.M. (1979). Growth and development of the pistillate flower and nut in 'Barcelona' filbert. *Journal of the American Society of Horticultural Science* **104**, 427–32.

Thomson, J.D. (1989). Germination schedules of pollen grains: Implications for pollen selection. *Evolution* **43**, 220–3.

Tilton, V.R. & Horner, H.T. (1980). Stigma, style and obturator of *Ornithogalum caudatum* (Liliaceae) and their function in the reproductive process. *American Journal of Botany* **67**, 1113–31.

Tilton, V.R., Wilcox, L.W., Palmer, R.G. & Albertsen, M.C. (1984). Stigma, style and obturator of soybean *Glycine max* (L.) Herr. (Leguminosae) and their function in the reproductive process. *American Journal of Botany* **71**, 676–86.

Tiwari, S.C. & Polito, V.S. (1988). Organization of the cytoskeleton in pollen tubes of *Pyrus communis*: A study employing conventional and freeze-substitution electron microscopy, immunofluorescence, and Rhodamine–Phalloidin. *Protoplasma* **147**, 100–12.

Van der Kloet, S.P. (1984). Effects of pollen donors on seed production, seed weight, germination, and seedling vigour in *Vaccinium corymbosum* L. *American Midland Naturalist* **112**, 392–6.

Van der Woude, W.J. & Morré, D.J. (1967). Endoplasmic reticulum-dictyosome–secretory vesicle associations in pollen tubes of *Lilium longiflorum* Thunb. *Proceedings of the Indiana Academy of Sciences* **77**, 164–70.

Vazart, J. (1971). Dégénérescence d'une synergide et pénétration du tube pollinique dans le sac embryonnaire du *Linum usitatissimum* L.

Annales de l'Université et de l'Association Régionale pour l'Etude de la Recherche Scientifiques (Reims) **9**, 133–9.

Webb, M.C. & Williams, E.G. (1988). The pollen tube pathway in the pistil of *Lycopersicon peruvianum*. *Annals of Botany* **61**, 415–23.

Welk, M., Millington, W.F. & Rosen, W.G. (1965). Chemotropic activity and the pathway of the pollen tube in lily. *American Journal of Botany* **52**, 774–81.

Went, J.L. van & Willemse, M.T.M. (1984). Fertilization. In *Embryology of the Angiosperms* (ed. B.M. Johri), pp. 273–317. Berlin: Springer-Verlag.

Went, J.L. van (1970). The ultrastructure of the synergids of *Petunia*. *Acta Botanica Neerlandica* **19**, 121–32.

Wetzstein, H.Y. & Sparks, D. (1989). Stigma–pollen interactions in Pecan. *Journal of the American Horticultural Society* **114**, 355–9.

Willemse, M.T.M. & Franssen-Verheijen, M.A.W. (1988). Pollen tube growth and its pathway in *Gasteria verrucosa* (Mill.) H. Duval. *Phytomorphology* **38**, 127–32.

Williams, E.G., Knox, R.B., Kaul, V. & Rouse, J.L. (1984). Post pollination callose development in the ovules of *Rhododendron* and *Ledum* (Ericaceae): zygote special wall. *Journal of Cell Science* **69**, 127–35.

Wilms, H.J. (1981). Pollen tube penetration and fertilization in spinach. *Acta Botanica Neerlandica* **30**, 101–22.

D. J. BOWLES

Embryogenesis

Introduction

In higher plants, fertilization of the ovule leads to the formation of the embryo. Zygotic embryogenesis, the way in which the fertilized ovule develops, has been studied extensively; changes that occur have been described for many plant species at the level of morphology, metabolism, protein composition, and gene expression. Interestingly, plant embryos can also develop in the absence of a fertilized ovule, from somatic cells in callus culture (Sung *et al.*, 1984), from differentiated cells such as leaf mesophyll (Conger *et al.*, 1983) and, perhaps most surprisingly, from immature haploid male gametes termed microspores (Nitsch, 1969; Dunwell, 1985). These alternative routes to the formation of an embryo illustrate both the means by which plants use environmental stimuli as developmental signals and the plasticity of plant development that is maintained throughout growth.

This article is not a comprehensive review of embryogenesis but rather a brief introduction to some interesting key issues in the area, and an outline of the ways in which we have been approaching the subject at Leeds. Two principal issues about plant embryogenesis will be discussed. The first concerns embryo formation, and the second concerns the way in which an embryo, fully capable of germinating within days of organ primordia differentiation, is nevertheless prevented from doing so until seed development has been completed.

Embryo formation

In theory, embryogenesis induced in culture should provide an ideal experimental system to analyse the factors that induce, determine and regulate embryo formation. For example, somatic embryogenesis in carrot suspension cultures has been studied extensively since its discovery in the 1950s (Reinert, 1958; Steward, Mapes & Hears, 1958), but in practice the system has proved to be more complex than first envisaged.

Current understanding of the carrot system

Studies have shown that cultures maintained in 2,4-D as proliferating cells could be converted to cultures containing embryos by the removal of 2,4-D from the medium (Nomura & Komamine, 1986). However, few changes in protein composition or translatable mRNAs were detected between the two populations, i.e. the genes expressed in the 'embryo' cultures appeared to be near-identical to those expressed in the 'non-embryo' cultures (Sung & Okimoto, 1981, 1983; Choi & Sung, 1984; de Vries et al., 1988).

It has since been suggested that the 'non-embryo' cultures were nevertheless 'embryonic' through the existence of structures initially described as proembryogenic masses (Halperin, 1966). The proembryogenic masses were found to be derived from a specific cell-type (Type 1 cells) under the influence of 2,4-D and were suggested to be stable intermediates in the developmental pathway linking single somatic cells to embryos (Nomura & Komamine, 1985). 2,4-D is therefore required to initiate embryogenic development, but inhibits further progress along the pathway at the stage of proembryogenic masses. Removal of 2,4-D was considered to release the embryogenic potential of the proembryogenic masses and lead to continuation of the pathway beyond the stage of the stable intermediate. Subsequently, by using an alternative means of producing proembryogenic masses, it has been shown that patterns of gene expression in the cell clumps are near-identical to those of heart-shaped (Thomas & Wilde, 1987) or torpedo-stage somatic embryos (Wilde et al., 1988).

These results highlight the problems of callus cultures, since although they are often assumed to consist of a homogeneous population of cell-types, in reality they are probably composed of a range of cell-types with a range of developmental potentials, each responding in different ways to the prevailing conditions of culture.

One of the interesting features of the carrot system is the relationship between vacuolation and the potential to become embryogenic. The proembryogenic masses were first detected through their typical morphology: small, rounded cells, dense cytoplasm and non-vacuolate (Halperin & Jensen, 1987). In the presence of 2,4-D, most of the carrot cells in culture had a different appearance, exhibiting typical patterns of cell extension growth and vacuolation. Highly vacuolate cells were not embryogenic, suggesting a prerequisite for induction of embryogenesis was the prevention of cell enlargement. Cell lines of carrot have been described that fail to form embryos when 2,4-D is removed from the culture medium (Lo Schiaco, Giuliano & Terzi, 1985). In one cell line that has been studied, the mutant cells were vacuolate: the mutants could be 'rescued' by addition of a glycoprotein to the culture that had been isolated from the medium of cells undergoing embryogenesis (de Vries, 1989). Rescue resulted in a changed morphology: the cells became rounded with dense cytoplasm. The glycoprotein was found to have peroxidase activity and it was suggested that the altered growth properties might arise from peroxidase-mediated cross-linking of wall polymers (e.g. isodityrosine cross-links) thereby preventing turgor-driven cell expansion. Do these results suggest, therefore, that to become embryos, cells must be restrained from expansion, and that once vacuolation has become initiated, embryogenesis is a non-starter?

The carrot system also highlights the profound influence of the extracellular matrix in determining the properties of the cells that the wall surrounds. Tunicamycin prevents *N*-glycosylation of proteins but, at sufficiently low concentrations, certain isomers of the antibiotic have little effect on protein synthesis (Elbein, 1987). Tunicamycin abolished the ability of carrot cells to become embryos, and the tunicamycin-treated cells could be rescued by the addition of extracellular glycoproteins from the medium of embryogenic cultures (de Vries *et al.*, 1988). On the assumption that *all* of the non-glycosylated proteins were still secreted in tunicamycin-treated cells, the results are surprising, because they imply that *N*-glycans may play a role in aquisition of embryogenic potential. Very recently, it has been found that addition of exogenous peroxidase to the cells could lead to partial rescue. This finding suggests that, at least in part, the tunicamycin effect was related to changes in secreted peroxidase. However, peroxidase activity in the medium of tunicamycin-treated cells was as high as in untreated cells, and since inactivated peroxidase (acetone-treated) could similarly rescue the situation, even more emphasis was placed on the role of the *N*-glycan(s) attached to the enzyme (S.C. de Vries, personal communication). Glycans of the cell wall are increasingly recognized to be important regulatory molecules both in defence gene activation and at some level in the determination of morphogenetic programming (Bowles, 1990). But the endogenous glycans identified to date have been fragments of polysaccharides (pectins or xyloglucans), not components of extracellular glycoproteins. Since *N*-glycosylation is a common occurrence in all eukaryotic cells, a regulatory role for *N*-glycans would have startling implications! Recent studies have also implicated extracellular arabinogalactan proteins as markers for cell identity during somatic embryogenesis in carrot (Stacey, Roberts & Knox, 1990). Withdrawal of 2,4-D from the medium led to a dramatic increase in expression of the epitope recognized by the monoclonal anti-body JIM 4. The epitope, a component of arabinogalactan proteins (Knox, Day & Roberts, 1989), was differentially expressed by cells at all stages from proembryogenic masses to mature embryos. This led to the suggestion that the arabinogalactan proteins identified may be function-ally concerned with the position of cells in relation to plant form.

Finally, the carrot system has also thrown into question the correlation between patterns of gene expression induced during embryogenesis *in vitro* and *in vivo*. This has arisen through the detection of a product encoded by a gene, Dc3 (Thomas & Wilde, 1987; Wilde *et al.*, 1988). A transcript complementary to Dc3 was found to be highly-expressed in globular and torpedo embryos, as well as proembryogenic masses (Wilde *et al.*, 1988; de Vries *et al.*, 1988). Because the gene was not expressed in hypocotyl nor leaf tissue of carrot, it was suggested to represent a marker for embryogenic potential. However, peptide sequence of the polypeptide, predicted from sequencing Dc3 cDNA, showed the protein could be classified as a late embryogenesis abundant (LEA) product, which by definition would not normally be expressed until late stages of development. LEAs have been suggested to be stress proteins, correlated with desiccation tolerance (Dure *et al.*, 1989).

Overall, therefore, the results suggest that the temporal pattern of gene expression of cultured embryos may not correlate with that of zygotic embryos, and the culture *per se* may lead to stress conditions not normally experienced by zygotic embryos *in planta* until much later during their maturation. This in turn leads to the idea that the control on morpho-genesis may be uncoupled from that on gene expression *per se*: two structures may appear morphologically identical, but express quite dif-ferent patterns of gene expression. Similarly, the genes required for maintaining an actively dividing cell are most probably not going to differ much between meristematic cells of the root, shoot, or young embryo, yet the eventual form of the structures derived from cell division in each case is clearly different.

The culture system in use at Leeds to study embryogenesis *in vitro*

In collaboration with the group of Dr Rob Lyne at Shell Research Ltd, Sittingbourne, we have investigated the route to embryogenesis that involves microspores: immature pollen. Anther culture leading to the direct formation of embryos from microspores has been established for a number of dicot species, such as tobacco (Raghavan, 1984). However, in cereals such as wheat or rice, the more typical route to embryogenesis from anther culture involves an intervening callus stage and the proliferation of secondary embryoids from the embryonic callus (Hu, 1986). In contrast, work at Sittingbourne has optimized a means of direct embryo formation from barley anther culture that does *not* involve callus (Lyne, Bennett & Hunter, 1984).

Within the defined parameters of the procedure, the culture system leads to the formation of well-formed, doubled haploid embryos at high efficiency (400 green plants per 100 anthers). In anther culture, proliferation is clearly visible by 14 d from plating out, and embryonic structures can be detected from 18 d onwards. Probably as a consequence of nutrient and growth-factor gradients within the anther (Hunter, 1985), production of embryos from microspores is not synchronous; a typical culture dish contains young embryos of different sizes as well as precociously germinated seedlings. Direct culture of microspores, i.e. removal from the anther before plating out, leads to a much more synchronous development of embryos (R.L. Lyne, unpublished results).

The similarity of embryos formed in culture and those formed *in planta*

We were interested to compare zygotic embryos of barley with those formed from microspores. This initially involved studies of patterns of translatable mRNA populations, in which we compared zygotic and microspore-derived embryos and barley callus (Higgins & Bowles, 1990). RNA extracted from the structures was translated *in vitro* within a rabbit reticulocyte lysate in the presence of [^{35}S]methionine, and the newly synthesized polypeptides were analysed by two-dimensional gel electrophoresis and visualized by autoradiography. The results indicated that the patterns of genes expressed in zygotic and microspore-derived embryos of comparable developmental stage were not identical. There were common features, but there were also very noticeable differences. In particular, a number of gene products found in the cultured embryos, but not in the zygotic embryos, were also detected in callus. Superficially, this result is similar to that obtained from carrot somatic embryogenesis, although the similarities in carrot embryos and callus were subsequently resolved by the detection of proembryogenic masses in the callus. The barley callus analysed in our studies was derived from scutellar tissue of embryos, but the callus had been subcultured over a period of several years before its use in the analyses. The similarities in the microspore-derived embryos and callus involved a range of gene products that were highly abundant, suggesting that they were not derived from only a small percentage of the cells analysed. An alternative explanation to the existence of proembryogenic cells in the barley callus is that the embryos formed in culture reflect that origin, through the expression of genes typical of cultured cells. Equally, it is possible that the gene products found in zygotic embryos, but not in cultured embryos, reflect the influence of the particular environment *in planta* within which the fertilized ovule develops. In these instances therefore, the gene products common to the two classes of embryo could provide more insight into what is required for the construction of an embryo *per se*, outside of the impact of the environmental conditions within which it develops.

Molecular markers for barley embryogenesis

We have used two approaches to identify gene products that can be used as molecular markers for embryogenesis: an immunological strategy, and one involving the construction and differential screening of a cDNA library. The projects are in relatively early stages; in the following discussion, results describing two 'markers', one produced by each approach, will be described. Conclusions from the work completed to date will then be outlined.

Characterization of an embryo-specific polypeptide of molecular mass 17 kDa

Extracts prepared from barley zygotic embryos (0.5 – 1.0 mm in length) were used for immunization. The antiserum was found to interact with a wide range of polypeptides on Western blots, many of which were common to embryos, leaf-base (meristematic tissue) and callus. However from these preliminary analyses, an immunoreactive polypeptide of molecular mass 17 kDa appeared to be located specifically in embryos (Higgins, 1989). An immunoaffinity system was used to prepare monospecific antibodies to the product, i.e. the antiserum was passaged through immobilized proteins extracted from leaf-base and callus. The effluent from the matrix was assayed on Western blots and contained antibodies that reacted only with the 17 kDa polypeptide (Clark *et al.*, 1991). The antigen is a low-abundance gene product, undetectable in SDS–PAGE profiles or total embryo extracts by either Coomassie blue or silver staining. Despite this high antigenicity, we can find no evidence for glycosylation, and the product does not cross-react with MAC 207, a monoclonal antibody that detects a carbohydrate determinant common to all arabinogalactan proteins (Pennell *et al.*, 1989). By using Triton X-114 partitioning, we find that the protein is amphiphilic in character. During zygotic embryogenesis, the product appears at mid-stages of development, at 18 d post-anthesis (dpa) (growth temperature 12 °C), and remains at similar levels up to 6 d post-germination of the seedling. The timing differs in cultured embryos: appearance of the polypeptide precedes the appearance of morphologically distinct embryos. Tissue-printing revealed that the antigen was distributed evenly throughout the embryo tissues, and within grains the polypeptide could also be detected in the aleurone layer but *not* in the endosperm.

The low abundance of the product, and the timing of its appearance and disappearance during zygotic embryo development and germination, suggests that it is not a LEA (Dure *et al.*, 1989). However, culture of young zygotic embryos on either ABA (10^{-6} M) or mannitol (9% w/v) led to the precocious appearance of the polypeptide. Preliminary analyses have shown that the protein is N-terminally blocked, but sequencing of peptide fragments produced by V-8 digestion is under way. We anticipate that the peptide sequence information gained will be used to design amplimers for polymerase chain reacton, thereby providing a way to obtain information on the gene encoding this embryo-specific product.

Temporal and spatial patterns of expression of pZE40

Because of the similarities in patterns of gene expression between microspore-derived embryos and callus, we decided to construct a cDNA library to mRNA from zygotic embryos in preference to embryos formed in culture. For the same reason, leaf-base material was chosen in preference to callus for screening out 'housekeeping' products. After three rounds of differential screening of the cDNA library, some 30 clones were detected, complementary to genes that were preferentially expressed in embryos compared with meristematic leaf-base tissue (Smith *et al.*, 1991). These clones are currently under investigation; the results concerning one, pZE40, will be introduced briefly here. In Northern analyses, a transcript of approx. 1.2 kb appears during early–mid stages of embryo development. The transcript is not present in dry mature embryos but reappears upon germination, 2–3 d after imbibition in water. By means of hybrid release translation, the transcript corresponding to pZE40 was found to encode a polypeptide of molecular mass approximately 40 kDa.

In situ hybridization was used to investigate the spatial expression of pZE40 and to extend the information on temporal regulation to very young embryos. Using *in situ* hybridization, we have detected pZE40 expression in embryos as young as 8 dpa. The gene is strongly expressed within the scutellum, coleorhiza and coleoptile of the embryo (i.e. the non-axial tissue) and the aleurone of the grain. Expression may also be detected in restricted cell populations at the primordial leaf tip.

Conclusions drawn from the two approaches

The problems of raising antisera to protein mixtures are notorious, with factors of immunodominance, the antigenicity of carbohydrate, etc., being particularly prevalent in studies of plant protein mixtures. With the protocols we used, the immunological approach led to the detection of only one 'embryo-specific' polypeptide of 17 kDa, which was not an abundant component of the antigen mixture but was clearly highly immunogenic. As yet we do not know the sequence of the protein, but its properties (timing of appearance–disappearance and localization) distinguish it from other products that have been described, such as the LEAs, barley germ agglutinin, and deterrent proteins such as the proteinase inhibitors that are generally expressed in the endosperm (Bowles, 1990). It is possible that it corresponds to a product of one of the genes identified recently in studies on barley aleurone, which were found to be expressed both in the aleurone *and* in the embryo of the grain (Jakobsen *et al.*, 1989). The relationship of the protein to previously identified products, and some insight into its function, may be forthcoming when we have sequence analysis.

Interestingly, the gene product identified in the alternative strategy involving cDNA library construction and screening, was also detectable in the aleurone as well as the embryo. However, this product, pZE40, was located specifically in the scutellum of the embryo, although not in the scutellar epithelium (the region of direct interface between the embryo and endosperm). As yet we have only partial sequence information on the product, but comparison with known sequences in the data banks has not revealed considerable homology.

One of the most notable features highlighted by the *in situ* hybridization studies is the subtlety in the spatial pattern of gene expression. Cells such as those constituting the population in the primordial leaf tips appear identical to those in their immediate environment, yet nevertheless express quite different genes. In turn, this must reflect a property unique to this small group of cells. Presumably, in relation to the expression of PZE40, this property is common to that of cells within the non-axial tissues of the embryo and the aleurone. Similar subtle distinction of cell-types has also been highlighted recently with the detection of differentiation antigens, extracellular markers of cell identity that predict future pattern formation at stages preceding any morphological criterion (Knox *et al.*, 1989). These examples and the lessons learnt from the carrot system, in particular those involving the embryonic potential of Type 1 cells, suggest that developmental decisions are probably taken much earlier than we usually realize.

Control of precocious germination

Once organ primordia have differentiated, the young immature embryo is capable of germination. This potential can be demonstrated in zygotic embryos through dissection of the embryo from the grain and culture *in vitro* (Norstog, 1972). Precocious germination occurs within several days, a process noted as early as 1904 by Hannig and subsequently described for a wide range of plant species. For embryos formed in culture, such as those derived from barley microspores, precocious germination occurs as a matter of course, once primordia have differentiated (Nitsch, 1969). Thus in culture, full maturation of the embryo does not occur and many genes characteristic of mid–late stages of embryo development are not expressed; for example, the gene encoding barley germ agglutinin is not expressed during embryogenesis in microspore culture (P.C. Morris & D.J. Bowles, unpublished results).

Since the zygotic embryo does not germinate precociously *in planta*, but does germinate immediately on removal from the plant, it follows that the environment surrounding the zygote influences the events that take place. The time period over which this germination potential is suppressed depends on varietal and environmental factors, but can be quite considerable. For example, in wheat or barley, the differentiated embryo is prevented from germinating for 50–60 d, the time taken to construct the desiccated, fully mature grain (Rogers & Quatrano, 1983).

During germination, growth occurs initially by rapid cell expansion following water uptake, and then by cell division within the meristems of the organ primordia. If this pattern of growth is initiated in precociously germinated immature embryos, it is clearly very different from that undertaken by the embryo that matures normally, in which all the cells are maintained at a similar size and shape.

The factors regulating the switch between embryogenic and germinative pathways of development have been studied in a number of laboratories world-wide, including research from my own group. In the following discussion, results obtained at Leeds on wheat and barley will be placed within the context of data from other laboratories.

The relationship between abscisic acid and osmotic stress

Culture of immature zygotic embryos on nutrient media leads to precocious germination. Inclusion of abscisic acid, or an osmotic agent such as mannitol, in the medium leads to the inhibition of precocious germination (Morris & Bowles, 1987). It seems highly relevant that the *concentration* of these two agents must be progressively increased to prevent the germination of progressively 'older' immature embryos. This phenomenon has been shown for both dicots and monocots (see, for example, Morris *et al.*, 1985; Finkelstein & Crouch, 1986). For example, abscisic acid at 10^{-6} M or mannitol at 4.5% (w/v) is sufficient to suppress precocious germination of young barley or wheat embryos (0.2 – 0.7 mg fresh mass), but 10^{-4} M abscisic acid and 9% (w/v) mannitol is required for 'older' embryos, e.g. 1.5 mg fresh mass (Morris *et al.*, 1985). This is not due to problems of uptake, since embryos on ABA-containing media take up the growth regulator to the concentration supplied (Walker-Simmons, 1987; Morris *et al.*, 1990).

Interestingly, studies of Schopfer & Plachy (1984, 1985), on prevention of germination in imbibed *Brassica* seeds, found the effects of exogenous ABA and osmotic agent to be *additive*, i.e. a low concentration of growth regulator could be compensated for by a higher concentration of osmotic agent and *vice versa*. The effect of exogenous ABA was found to be on the control of cell wall extensibility and the minimum threshold turgor pressure for expansion. A prerequisite for turgor-driven cell expansion is the possibility that polymers within the extracellular matrix can slide past one another to accommodate the expansion of the protoplast. Exogenous ABA could very readily affect apoplastic pH: it is known, for example, to inhibit the action of acid hydrolases in the cell wall, and its effect is reversible by the application of fusicoccin (Labrador, Rodriguez & Nicolas, 1987), a toxin known to affect proton transport. Thus, from the work on *Brassica napus*, it is possible to envisage that exogenous ABA inhibits germination by affecting cell wall extensibility, whereas high external osmotica would affect expansion simply by preventing internal turgor pressure.

If we return to the problem of suppression of germination potential during embryo development, it is equally possible that ABA and the water relations of the embryo and grain, play two distinct roles in the process. This is an alternative explanation to the one more generally considered, namely that endogenous ABA is the sole direct mediator of the switch between continued embryo development and precocious germination. This latter viewpoint suggests that osmotic stress may be the environmental stimulus in culture or *in planta*, but the transduction pathway linking that stimulus to gene activation or suppression involves ABA. Measurement of endogenous ABA in cultured embryos of dicots (Finkelstein *et al.*, 1985; Finkelstein & Crouch, 1986; Eisenberg & Mascarenhas, 1985), or cereals (Morris *et al.*, 1988; Walker-Simmons, 1987) suggests that there is more to the story than a simple increase of internal ABA in response to stress imposed by the external osmoticum.

Indeed, in *Brassica napus* embryos or in cereal embryos, culture on high osmoticum does not necessarily lead to an increased level of endogenous ABA. Also, ABA-insensitive mutants of maize can be prevented from precocious germination by culture on high osmoticum, whereas exogenous ABA has no effect (A.C. Cumming, personal communication). The growth regulator may well be the direct signal for activation of some genes however, since, for example, incubation of transformed rice protoplasts in ABA was found to activate very rapidly a wheat E_m gene, which is known to be regulated in cultured zygotic embryos by either exogenous ABA or osmotica (Marcotte, Bagley & Quatrano, 1988).

Suppression of germination *in planta*

A number of studies have determined the changes in level of ABA during seed development, for example wheat and barley (McWha, 1975; Quarrie, Tuberosa & Lister, 1988; Quarrie *et al.*, 1988), soybean (Akerson, 1984), and *Brassica napus* (Finkelstein *et al.*, 1985). The precise pattern of change is dependent on species, genotype and environment, but follows a general trend: ABA levels steadily increase during the early stages of development to a maximum value and then decline as the seed dehydrates. Viviparous mutants of maize have provided good evidence for an involvement of ABA in control of vivipary, since embryos from the mutant plants have been shown to be insensitive to exogenous ABA (McDaniel, Smith & Price, 1977) or to contain abnormally low levels of the growth regulator (Brenner, Burr & Burr, 1977).

Recently, we have followed a number of parameters in parallel during cereal embryo development *in planta*, in order to examine endogenous ABA levels within the context of the water relations of the embryo and the grain. Thus, ABA levels, fresh mass, dry mass, and water potential were assayed for embryos and for the remainder of the grain, at intervals from shortly after anthesis through to desiccation and full maturation at 65 dpa (Morris *et al.*, 1990).

The overall concentration of ABA, determined by immunoassay using standardized procedures (Morris *et al.*, 1988), fluctuated only 0.25 – 2.0 pmol mg^{-1} fresh mass during development, whether in the grain or embryo, and followed the general trend described in earlier studies. Thus, at early stages of embryo formation, precocious germination could in practice be prevented by the endogenous levels of ABA, since they correspond to the values known to inhibit the germination of young, immature cultured embryos (e.g. 0.2 mg fresh mass). However, by later stages of development, for example by the time embryos are 1.5 mg fresh mass, it is known that 50–100 μM exogenous ABA is required to prevent their precocious germination in culture. This value clearly exceeds the level found in embryos, or grains from the plant.

Interestingly, as development proceeded from 18–20 dpa onwards, a progressive increase in water potential difference was set up between the embryo and the rest of the grain. Thus, at the earliest time-point measured (2 dpa), the grain (at this stage, consisting of a coenocyticum with very high levels of low-molecular-mass metabolites) had a negative water potential of −5 to −6 MPa. By 7 dpa, the water potential changed to −3 to −2 MPa, perhaps reflecting a decreasing osmotic potential through conversion of metabolites to polymers and an increasing turgor pressure through the deposition of cell walls within the endosperm. This value of water potential for the grain was then maintained through to 21 dpa, and was found to be near identical to the water potential of the embryo. However, from 21 to 50 dpa, the water potential of the embryo remained constant, whereas that of the rest of the grain decreased steadily to −8 MPa. This therefore led to an increasing water potential difference between the embryo and the grain, such that cell expansion of the developing embryo would be negligible through lack of water uptake. Only by very late stages of maturation, at 50 dpa, did the water potential of the embryo again resemble that of the rest of the grain, when values changed from −2 to −8 MPa in less than 7 days and the seed approached full desiccation.

These results suggest that, *in planta*, both ABA and the water relations of the system may control the switch between continued embryo development or germination. At the early stages, endogenous ABA levels are those predicted from *in vitro* studies as sufficient to suppress germination potential. By mid-stages of maturation, the water potential difference that is progressively established between the embryo and the rest of the grain is sufficient to ensure that cell expansion is inhibited by prevention of water uptake. It is possible that, *in vitro*, the unphysiological high levels of exogenous ABA required to prevent germination of the larger embryos act in a manner suggested by Schopfer's work, i.e. indirectly mimic the *in planta* control, through blocking cell-wall extensibility and thereby inhibiting water uptake.

Final comments

The consequence of a successful pathway of embryogenesis is a structure capable of germination. Despite the fact that these structures can be induced *in planta* or *in vitro*, it would seem to be comparatively rare that research workers in the two fields converse. Somatic embryogenesis is regarded as an issue of tissue culture, whereas zygotic embryogenesis is regarded as an issue of seed and fruit physiology and ecology, yet both routes to the formation of an embryo involve morphologically identical structures that exhibit at least similar patterns of gene expression. Perhaps with an increasingly inter-disciplinary approach to plant science, we will be able to combine molecular and cellular information and more rapidly understand the signals acting on cells that lead to the construction of a unique three-dimensional structure such as a plant embryo.

Acknowledgements

The research on cereal embryogenesis has been supported by two SERC–CASE studentships, a SERC grant and a SERC Co-Operative grant with Shell Research Ltd, Sittingbourne, to DJB. Studies on the role of ABA and osmotic potential in wheat and barley involved Dr Peter Morris and Dr Peter Jewer. Studies on zygotic embryogenesis and microspore-derived embryogenesis in barley involved Anthony Clark, Linda Donovan, Dr Trish Higgins, Helen Martin, Dr Laura Smith, Jane Handley and Li Yi.

References

Ackerson, R.C. (1984). Regulation of soybean embryogenesis by abscisic acid. *Journal of Experimental Botany* **35**, 403–13.

Bowles, D.J. (1990). Defense-related proteins in higher plants. *Annual Review of Biochemistry* **59**, 873–907.

Brenner, M.L., Burr, B. & Burr, F. (1977). Correlation of genetic vivipary in corn with abscisic acid concentration. *Plant Physiology* **59**, 76.

Choi, J.H. & Sung, Z.R. (1984). Two-dimensional gel analysis of carrot somatic embryonic proteins. *Plant Molecular Biology Reporter* **2**, 19–25.

Clark, A.J., Higgins, P., Martin, H. & Bowles, D.J. (1991). An embryo-specific protein of barley. *European Journal of Biochemistry* **199**, 115–21.

Conger, B.V., Hanning, G.E., Gray, D.J. & McDaniel, J.K. (1983). Direct embryogenesis from mesophyll cells of orchard grass. *Science* **211**, 850–1.

Dunwell, J.M. (1985). Embryogenesis from pollen in vitro. In *Biotechnology in Plant Science – Relevance to Agriculture in the Eighties* (ed. M. Aitlin, P. Day & A. Hollander), pp. 49–76. London: Academic Press.

Dure, L.S., Crouch, M., Harada, J., Ho, T.H., Mundy, J., Quatrano, R.S., Thomas, T. & Sung, Z.R. (1989). Common amino acid sequence domains among the LEA proteins of higher plants. *Plant Molecular Biology* **12**, 475–86.

Eisenberg, A.J. & Mascarenhas, J.P. (1985). Abscisic acid and the regulation of synthesis of specific seed proteins and their messenger RNAs during culture of soybean embryos. *Planta* **166**, 505–14.

Elbein, A.D. (1987). Inhibitors of the biosynthesis and processing of N-linked oligosaccharide chains. *Annual Review of Biochemistry* **56**, 497–534.

Finkenstein, R.R. & Crouch, M.L. (1986). Rape seed embryo development in culture on high osmoticum is similar to that in seeds. *Plant Physiology* **81**, 907–12.

Finkelstein, R.R., Tenbarge, K.M., Shumway, J.E. & Crouch, M.L. (1985). Role of ABA in maturation of rapeseed embryos. *Plant Physiology* **78**, 630–6.

Halperin, W. (1966). Alternative morphogenetic events in cell suspensions. *American Journal of Botany* **53**, 443–53.

Halperin, W. & Jenson, W.A. (1987). Ultrastructural changes during growth and embryogenesis in carrot cell cultures. *Journal of Ultrastructural Research* **18**, 428–43.

Hannig, E. (1904). Zur Physiologie pflanzlicher embryonen. *Botanische Zeitschrift* **62**, 45–80.

Higgins, P. (1989). *Embryogenesis in barley anther culture*. Ph.D. thesis, University of Leeds.

Higgins, P. & Bowles, D.J. (1990). Comparative analysis of translatable mRNA populations in zygotic and pollen-derived embryos of barley *Hordeum vulgare*. *Plant Science* **69**, 239–47.

Hu, H. (1986). Wheat improvement through anther culture in biotechnology. In *Agriculture and Forestry 2. Crops 1* (ed. Y.P.S. Bajaj), pp. 55–72. Berlin: Springer-Verlag.

Hunter, C.P. (1985). The effect of anther orientation on the production of microspore-derived and plants of *Hordeum vulgare*. *Plant Cell Report* **4**, 267–8.

Jakobsen, K., Klemsdal, S.S., Aalen, R.B., Bosnes, M., Alexander, D. & Olsen, O.A. (1989). Barley aleurone cell development: molecular cloning of aleurone-specific cDNAs from immature grains. *Plant Molecular Biology* **12**, 285–93.

Knox, J.P., Day, S. & Roberts, K. (1989). A set of cell surface glycoproteins forms an early marker of cell position and not cell-type, in root apical meristem of *Daucus carota*. *Development* **106**, 47–56.

Labrador, E., Rodriguez, D. & Nicolas, G. (1987). Changes in cell wall composition of embryonic axes of germinating *Cicer arietinum* with temperature. *Plant Science* **48**, 23–30.

Lyne, R.L., Bennett, R.I. & Hunter, C.P. (1984). Embryoid and plant production from cultured barley anthers. In *Plant Tissue Culture and its Agricultural Applications* (ed. L.A. Withers & P.G. Anderson), pp. 405–11. London: Butterworths.

Lo Schiaco, F., Giuliano, G. & Terzi, M. (1985). Pattern of polypeptides secreted in the conditioned medium and its alterations in a mutant TS for embryogenesis. In *Proceedings of Workshop on Somatic Embryogenesis* (ed. M. Terzi, L. Pitto & Z.R. Sung), pp. 32–4 : EPRA.

McDaniel, S., Smith, J.D. & Price, H.J. (1977). Response of viviparous mutants to abscisic acid. *Maize Genetics Newsletter* **51**, 85–6.

McWha, J.A. (1975). Changes in abscisic acid levels in developing grains of wheat. *Journal of Experimental Botany* **26**, 823–7.

Marcotte, W.R., Bagley, C.C. & Quatrano, R.S. (1988). Regulation of a wheat promotor by abscisic acid in rice protoplasts. *Nature* **335**, 454–7.

Morris, P.C. & Bowles, D.J. (1987). Abscisic acid and embryogenesis. In *Growth Regulators and Seeds* (ed. N.G. Pinfield & M. Black), pp. 65–76. Long Ashton, England: British Plant Growth Regulator Group.

Morris, P.C., Kumar, A., Bowles, D.J. & Cuming, A.C. (1990). Osmotic stress and abscisic acid induce expression of the wheat E_m genes. *European Journal of Biochemistry* **190**, 625–30.

Morris, P.C., Maddock, S.E., Jones, M.G.K. & Bowles, D.J. (1985). Changes in the levels of wheat- and barley-germ agglutinin during embryogenesis *in vivo, in vitro* and during germinations. *Planta* **166**, 407–13.

Morris, P.C., Weiler, E.W., Maddock, S.E., Jones, M.G.K., Lenton, J.R. & Bowles, D.J. (1988). Determination of endogenous ABA levels in immature cereal embryos during *in vitro* culture. *Planta* **173**, 110–16.

Nitsch, J.P. (1969). Experimental androgenesis in *Nicotiana. Phytomorphology* **19**, 389–404.

Nomura, K. & Komamine, A. (1985). Identification and isolation of single cells that produce somatic embryos at a high frequency in carrot suspension culture. *Plant Physiology* **79**, 988–91.

Nomura, K. & Komamine, A. (1986). Molecular mechanisms of somatic embryogenesis. *Oxford Surveys, Plant Molecular Cell Biology* **3**, 456–66.

Norstog, K. (1972). Factors relating to precocious germination in cultured barley embryos. *Phytomorphology* **22**, 134–9.

Pennell, R.I., Knox, J.P., Scofield, G.N., Selvendran, R.R. & Roberts, K. (1989). A family of abundant plasma membrane-associated glycoproteins related to the arabinogalactan proteins is unique to flowering plants. *Journal of Cell Biology* **108**, 1967–77.

Quarrie, S.A., Tuberosa, R. & Lister, P.G. (1988). Abscisic acid in developing grains of wheat and barley genotypes differing in grain weight. *Plant Growth Regulation* **7**, 3–17.

Quarrie, S.A., Whitford, P.N., Appleford, N.E.J., Wang, T.L., Cook, S.K., Henson, I.E. & Loveys, B.R. (1988). A monoclonal antibody to abscisic acid: its characterisation and use in a radioimmunoassay for measuring abscisic acid in crude extracts of cereal and lupin leaves. *Planta* **173**, 303–39.

Raghavan, V. (1984). Biochemistry of Somatic Embryogenesis. In *Handbook of Plant Cell Culture*, vol. 1 (ed. D.A. Evans, W.R. Sharp, P.V. Ammirato & Y. Yamada), pp. 654–71. London: Macmillan.

Reinert, J. (1958). Morphogenese und ihre Kontrolle an Gewebekulturen ans Carotten. *Naturwissenschaften* **45**, 334–45.

Rogers, S.O. & Quatrano, R.S. (1983). Morphological staging of wheat caryopsis development. *American Journal of Botany* **70**, 308–11.

Schopfer, P. & Plachy, C. (1984). Control of seed germination by abscisic acid. II. Effect on embryo water uptake in *Brassica napus*. *Plant Physiology* **76**, 155–60.

Schopfer, P. & Plachy, C. (1985). Control of seed germination by abscisic acid. III. Effect on embryo growth potential and growth coefficient in *Brassica napus*. *Plant Physiology* **77**, 676–86.

Smith, L.M., Handley, J., Yi, L., Martin, H. & Bowles, D.J. (1991). Temporal and spatial regulation of a novel gene in barley embryos. *Molecular and General Genetics* (submitted).

Stacey, N.J., Roberts, K. & Knox, J.P. (1990). Patterns of expression of the JIM 4 arabinogalactan-protein epitope in cell cultures during somatic embryogenesis of *Daucus carota*. *Planta* **180**, 285–92.

Steward, F.C., Mapes, M.D. & Hears, K. (1958). Growth and organized development of cultured cells. *American Journal of Botany* **45**, 705–8.

Sung, Z.R., Feingberg, A., Chorneau, R., Borkird, C., Furner, I., Smith, J., Terzi, M., Schaviavo, F., Giuliano, G., Pitto, L. & Nuti-Ronchi, V. (1984). Developmental biology of embryogenesis from carrot culture. *Plant Molecular Biology Reporter* **2**, 3–14.

Sung, Z.R. & Okimoto, R. (1981). Embryonic proteins in somatic embryos of carrot. *Proceedings of the National Academy of Sciences, USA* **78**, 3683–7.

Sung, Z.R. & Okimoto, R. (1983). Co-ordinate gene expression during somatic embryogenesis in carrot. *Proceedings of the National Academy of Sciences, USA* **80**, 2661–5.

Thomas, T.L. & Wilde, D. (1987). Analysis of carrot somatic embryo programmes. In *Proceedings of VI International Congress on Plant Cell and Tissue Culture*, pp. 83–93. New York: Alan R. Liss.

de Vries, S.C., Booij, H., Janssens, R., Vogels, R., Saris, L., Lo Schiavo, F., Terzi, M. & van Kammen, A. (1988). Carrot somatic embryogenesis depends on the phytohormone-controlled presence of correctly glycosylated extracellular proteins. *Genes and Development* **2**, 462–76.

de Vries, S.C., Booig, H., Meyerink, P., Huisman, G., Wilde, H.D.,

Thomas, T.L. & van Kammen, A. (1988). Aquisition of embryogenic potential in carrot suspension cultures. *Planta* **176**, 196–204.

de Vries, S.C. (1989). Developmental mutants and extracellular proteins in carrot somatic embryogenesis. In *Proceedings of 15th EMBO Annual Symposium.*

Walker-Simmons, M.L. (1987). ABA levels and sensitivity in developing wheat embryos of sprouting resistant and susceptible cultivars. *Plant Physiology* **84**, 61–6.

Wilde, H.D., Nelson, W.S., Booij, H., de Vries, S.C. & Thomas, T.L. (1988). Gene expression patterns in embryogenic and non-embryogenic carrot cultures. *Planta* **176**, 205–11.

G.K. GOLDWIN

Environmental and internal regulation of fruiting, with particular reference to Cox's Orange Pippin apple

Introduction

In many species, there is a long period from flower induction to fruit ripening and abscission, during which environmental factors can have a major impact on the success of fruiting. This is particularly so with perennial crops such as pome, stone and citrus fruits, where even the period just from anthesis to ripening can last up to 60 weeks (Valencia orange). It is not surprising, therefore, that there are several opportunities for adverse environmental conditions to lead to reduced fruit numbers at harvest. For example, flower abortion before anthesis can result from low light levels (e.g. tomato) (Kinet *et al.*, 1978), frost damage (e.g. pome fruits) (Modlibowska, 1964), water deficit (e.g. *Vicia faba* L.) (El-Nadi, 1969) and water excess (e.g. *Vicia faba* L.) (Smith, 1982), while shading can precipitate the drop of enlarging fruits (e.g. peach and apple) (Byers *et al.*, 1985) and wind can exacerbate the pre-harvest drop of pome fruits. Flower and fruit abortion has been reviewed by Stephenson (1981) and will not be the major topic of this chapter.

Although environmental factors can affect fruiting pre-anthesis and after the initial set has taken place, by far the most important period determining the success of fruiting comprises the few days during and after flower opening, when the transition from flower to fruitlet takes place. This, of course, is not surprising since it is also the time when the reproductive competitiveness of the maternal genome is partly determined through the success of fertilization and subsequent seed dispersal. Fruit development, though, is not always associated with ovule fertilization: since parthenocarpy can be stimulated in some species by frost damage (e.g. pear) (Lewis, 1942; Modlibowska, 1963), by pollination alone (e.g. tobacco (Yasuda, Inaba & Takahashi, 1935) and *Melandrium* (Van Overbeek, Conklin & Blakeslee, 1941)), or artificially by the application of plant hormones. Parthenocarpy has even been recorded when no setting stimulus was applied (neither pollen nor hormone), by simply altering the competitive balance between vegetative and reproductive structures (e.g. *Vicia faba* L.) (Chapman, Fagg & Peat, 1979).

Several reviews of fruit development and setting have been published; one, edited by Monselise (1986), describes in detail the growth of 26 fruits from set to harvesting. Coombe (1976) has described the origins, structure and development of fleshy fruits; the hormonal factors involved in fruit set and development have been reviewed by Crane (1964), Nitsch (1970), Nitsch (1971), Moore & Ecklund (1975), Goodwin (1978), Naylor (1984) and Treharne *et al.* (1985). As Browning (1989) has recently critically assessed the evidence for endogenous hormonal regulation of fruit set and development, this aspect will not be considered further here; emphasis will be placed more on the exogenous hormonal and environmental control of fruiting, with special reference to Cox's Orange Pippin apple.

Regulation of fruiting by effects on flower numbers

The first opportunity that the environment has in influencing fruit numbers occurs at the time of flower initiation. With many tree fruits, flowers are initiated in the summer and autumn previous to flower opening. With Cox apple, axillary buds need to initiate about 20 vegetative primordia including bud scales before the first flower primordium is produced (Abbott, 1977). In the UK, this is usually achieved in July at the earliest, but flower initiation can continue through to late autumn depending on the weather, and even recommence early in the following spring (Zeller, 1960). Warm, sunny days favour flower initiation and would be expected to lead to high flower numbers the following spring. The mechanism underlying this regulatory process is thought to relate to the production of photosynthate and the effect of temperature on the rate of primordial initiation. Abbott (1973) found, by using controlled-environment rooms, that higher temperatures in the period from August to December favoured flower bud initiation.

The likely importance of carbohydrate availability is highlighted by the inhibitory effect of shading on floral bud initiation (Jackson & Palmer, 1977) and the internal regulation of flowering operating through the presence of fruit on the tree. Williams *et al.* (1980) have shown that the time of harvesting Bramley's Seedling apple affects the number and quality of flowers the following spring, early harvesting favouring flowering and cropping. The authors suggested that early fruit removal increased node formation and advanced flower initiation. In addition, it was postulated that it enhanced carbohydrate accumulation and resulted in greater ovule longevity the following spring. The prevention of cropping by early fruitlet removal has an even greater promotory effect on flowering (Buszard & Schwabe, 1989; Vemmos, 1990), providing that excessive vegetative growth is not stimulated.

The alternative explanation, that fruit removal also removes seeds, from which substances inhibitory to flower initiation (e.g. gibberellins) have been shown to diffuse (Hoad, 1978), must also be considered. The hypothesis of seed inhibition of flowering is supported by the work of Tromp (1982), who found that the flowering of young apple trees was inhibited by the application of gibberellins A_3 and A_7 during flowering in the *previous* year. Because this timing is well in advance of the time that flower primordia are first observed, it must be concluded that gibberellin inhibited flower 'evocation'. Silva, Holgate & Abbott (1980) have reported that, with Golden Delicious apples, even the presence of flowers is inhibitory to the initiation of flowers in the following spring. It thus appears that fertilized or, to a lesser extent, unfertilized ovules can indirectly inhibit fruiting in the following year. This hypothesis is supported by the observations of the author that flowering and fruiting of monoculture Cox apple orchards, in which fruit are produced with few or no seeds, are more regular than in conventionally cross-pollinated orchards where fruit are produced with a higher seed content (Goldwin, 1986).

Regulation of fruiting by effects on initial fruit set

Fruit set is a rather loose term, which has often been used to refer to the success of flowers in reaching harvestable fruit. This definition, however, conceals the three obstacles that the flower has to negotiate successfully. Firstly, the slowing of ovary growth that occurs at flower opening must be reversed, and abscission at the base of the pedicel must be prevented. Secondly, the enlarging fruitlet must survive the intense period of competition from other reproductive and vegetative organs, and thirdly, environmental factors such as adverse weather conditions (e.g. high winds) and pest predation can remove the fruit just as it is reaching maturity. Experimentally, it is common to allow for these three distinct times when fruit numbers can be limited, by the terminology: initial set, final set and harvested set. Initial set can be defined as the proportion of flowers in which the ovary is stimulated to recommence rapid growth after anthesis. Final set can be defined as the proportion of flowers which survive the period of competition (in tree fruits often termed 'June drop'); harvested set can be defined as the proportion of flowers that are harvested as fruit.

Effects on pollination

Common events that occur at flowering in fruit species are petal opening, anthesis, nectar production, stigmatic secretion, pollen germination, pollen tube growth, embryo sac maturation, fusion of polar nuclei, double fertilization, petal fall and, if initial set is successful, the establishment of rapid ovary growth. The timing at which each of these events will occur depends largely on temperature; low temperatures tend to slow the pace of events, sometimes to such an extent that initial fruit set is unsuccessful. When initial set is stimulated by pollination, environmental factors can limit the process in several different ways, depending partly on whether incompatiblity mechanisms are inhibiting pollen tube growth in the style. Gustafson (1937) described earlier reports of experiments using dead or foreign pollen and concluded that pollen contained a substance promotory to ovary growth and that just the presence of pollen on the stigma could stimulate setting. However, this effect does not occur with all fruiting species. Nevertheless, regardless of the mode of stimulation, the transfer of pollen to the stigma is subject to environmental factors, particularly where the transfer is insect dependent. Bees, a common pollen vector, are greatly influenced by weather. Thus honey bees (*Apis mellifera*), often used by fruit growers to promote pollen transfer, are reluctant to fly when the temperature is low, when the weather is cloudy, or in high wind (Free, 1970). If these conditions prevail throughout the flowering period, then poor bee pollination is likely to result. Other pollinating insects such as some solitary and bumble bees are not so limited in their weather requirements (Free, 1970).

Once the pollen has alighted on the stigma, germination takes place and this event is, in itself, affected by environmental conditions (see Owens, this volume). In plum cultivars, the speed of germination was linearly related to temperature (Keulemans & van Laer, 1989). Whether an individual pollen tube will grow from stigma to embryo sac will depend firstly on whether the pollen is incompatible (in which case growth will usually stop close to the distal end of the style), secondly on whether other pollen tubes are present, and thirdly on whether growth is fast enough to enable the pollen tube to reach the base of the style before degeneration of the style has occurred. Incompatibility can be overcome by the presence of mentor pollen, as demonstrated with the successful selfing of Doyenné du Comice pear (Visser & Oost, 1982). Even when the pollen and style are compatible, the growth rates of pollen tubes can vary depending on the species and cultivars of the style and pollen (Modlibowska, 1945). Furthermore, pollen-tube growth is highly temperature-dependent (Modlibowska, 1945; Williams, 1970; Keulemans

& van Laer, 1989). For example, Williams (1970) found with apple that there was a linear relationship between temperature and the time at which fertilization took place after pollination, and Keulemans & van Laer (1989) showed that the speed of plum pollen-tube growth *in vitro* was not only linearly related to temperature, but varied markedly with cultivar. The effect that these factors controlling pollen-tube growth have on initial set depends on how far down the style the tubes need to reach before a signal can be transmitted to the ovary to stimulate growth. There appears to be little evidence in the literature regarding this point. Fuller & Leopold (1976) reported that the signal for fruit set in cucumber reached the ovary at about the time the pollen tubes entered the ovary. Initial examination by the author of results from an experiment in which styles were removed on successive days after hand-pollination of Cox apple flowers suggests that the pollen tubes need to travel a long way down the path to the embryo sacs before receptacle growth is stimulated in this cultivar. However, an histological examination of the flowers needs to be completed before this point can be clarified. Clearly, if the signal can be given by pollen tubes soon after entering the style, then fruitlet growth might be stimulated in some species by incompatible pollen. This would not be advantageous to the plant as fruits could then be produced without viable seeds, hence devoting resources to unprofitable reproductive development.

In contrast, if there is a requirement for pollen tubes to reach the embryo sac, or close to the embryo sac, before rapid ovary growth is re-established, then there is a much greater chance that the ovules of ovary tissues will degenerate before fertilization is effected. Herrero & Gascon (1987) have reported that in pear, pollination delays embryo sac degeneration, but very recent evidence obtained by G.K. Goldwin & M. Herrero (unpublished data) has shown that this effect is mediated through receptacle growth, since the delay only occurs in growing ovaries. Thus, it appears that, provided ovary growth can be stimulated, then embryo-sac longevity is extended and hence the probability of fertilization is increased.

Effects on the initial set of unpollinated flowers

Although pollination is by far the most commonly used stimulus for fruit set, parthenocarpy has been induced experimentally in a wide range of fruits, as summarized by Schwabe & Mills (1981). The attraction of achieving regular fruit cropping unrestrained by weather has stimulated considerable interest over many years in the practical use of plant hormones for setting. However, the high expectations have not been fulfilled with some species, partly because the underlying physiology of fruit setting was (and is still) not fully understood, and partly for practical reasons, such as the high production costs of some plant hormones and the very high costs associated with clearing a new chemical treatment for the market (Lever, 1985). A further practical factor that has not always been considered is the inverse relation between fruit numbers and fruit size. With the increasing market requirement for larger fruit, treatments that greatly increase fruit numbers can reduce the value of the crop. Despite these problems, plant hormones are currently being used to improve the fruit set of citrus, cherry, grape, apple, pear and tomato.

For the plant physiologist, the induction of parthenocarpy is a very convenient model system for studying fruit setting because the setting stimulus can be readily standardized in quantity, site of application and timing. Furthermore, the technique eliminates the involvement of the male gametes with all the inherent uncertainties in pollen germination and pollen-tube growth down the style.

It has long been recognized (Howlett, 1927) that not all flowers are capable of setting, even when hand-pollination has been carried out under ideal conditions. That differences in the setting potential of flowers of the same cultivar can exist between years has been accepted by fruit growers, but a precise investigation of the reasons for differences in flower 'quality' has not been easy when using pollination as the setting stimulus, because set failure can arise from defects at several different sites in the flower (e.g. stigma, style, ovule or pedicel). Differences in flower 'quality', however, have also been observed in the setting response to hormone (Goldwin, 1978) where any defects relating to the style are irrelevant. By using hormone-induced parthenocarpy, it has been possible to demonstate that flower 'quality' varies not only between years, between plants of the same clone and between clusters on the same branch, but also between flowers within the same inflorescence (Goldwin, 1981). It is essential, therefore, that investigations of fruit set should use systems that are as standardized as possible. It has been customary for the apical or 'king' flower in the inflorescence to be removed during setting experiments with pome fruits, as the proportion of these flowers that set is often different from those of the axillary flowers in the cluster, even when set parthenocarpically (Goldwin, 1981). Furthermore, in apple, these flowers open in advance of the others, when the pollination conditions are often different from those prevailing later. In order to separate environmental factors from internal factors, it is also essential to use flower populations that are not only sited on the same-aged wood but also synchronized in their stage of development. These requirements have rarely been fulfilled in fruit-setting studies, and perhaps explain the confusing results that have sometimes been obtained.

There are several possible reasons why flowers given the same stimulus should differ in their ability to set initially. Firstly, competition might be responsible and act in several different ways. There could be competition for carbohydrate, water, essential nutrients or amino acids, or competition could take place through an hormonal mechanism in which abscission of some flowers might be promoted by the hormonal efflux from others. Secondly, flowers might fail to set owing to a defect in the mechanism which re-establishes rapid ovary growth after anthesis. Thirdly, flowers might fail owing to a poorly developed vascular system.

Initial set limited by competition

Competition has been studied experimentally by manipulating the natural balance between reproductive organs or between reproductive and vegetative organs, or by restricting the availability of substrate (e.g. leaf removal or shading to reduce the carbohydrate pool). For example, the removal of the vegetative apex of *Vicia faba* plants increased the number of pods set and enabled unpollinated flowers to develop pods (Chapman *et al.*, 1979). Despite this, not all flowers set; competition for resources between pods was clearly seen in the pod distribution of different inflorescences.

Experiments carried out by the author with Cox apple flowers have also examined between-flower competition during initial set. For example, in a flower-thinning study conducted at Wye, sixteen spur clusters were selected on each of 25 Cox's Orange Pippin apple trees. The 'king' flower was removed from each cluster. Clusters were reduced to one, two, three or four flowers at the balloon stage; the flowers were then decapitated to prevent pollination. A 'Wye' hormone mixture, containing 200 mg l^{-1} gibberellic acid (GA$_3$), 50 mg l^{-1} 2-naphthyloxyacetic acid (NOXA) and 300 mg l^{-1} N,N'-diphenylurea (DPU) was applied to the flowers when they would have been at full bloom and the initial and final sets of parthenocarpic fruitlets recorded subsequently. The results are shown in Fig. 1; it is evident that the initial sets of flowers that were in clusters on their own or in pairs, were very similar (Fig. 1*a*). However, the mean initial set decreased where there were three or four flowers present, but the number of fruitlets set per cluster was still highest where four flowers were initially present (Fig. 1*b*). 'June drop' was heaviest in clusters originally thinned to three or four flowers, and the final set of flowers, originally four to a cluster, was less than half that of single flowers (Fig. 1*a*). The final number of fruit per cluster was similar for clusters that originally started with two, three or four flowers (Fig. 1*b*).

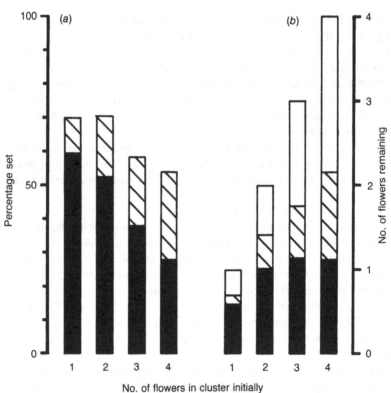

Fig. 1. (a) The mean sets (%) of hormone-treated, unpollinated Cox flowers thinned to 1, 2, 3 or 4 flowers per cluster. Hatched bars, initial set; solid bars, final set. (b) The mean numbers of flowers initially (open bars), numbers of fruitlets after initial set (hatched bars) and numbers of fruits after 'June drop' (solid bars) in clusters thinned to 1, 2, 3 or 4 flowers.

These results showed that competition between unpollinated Cox flowers during the period of initial set was relatively low when three or four flowers were present, and not apparent with only two. In earlier work with Cox clusters thinned to four flowers (Goldwin, 1984), competition between hormone-treated, unpollinated flowers and open-pollinated flowers in the same cluster was investigated. In one year, the initial set of parthenocarpic fruitlets was significantly reduced by the presence of open-pollinated fruitlets, the set of which was about six times that of the parthenocarpic fruitlets. However, in another year, there was no difference between the initial set of the unpollinated flower in the cluster and that of the three, remaining, open-pollinated flowers. In both years, the sets of hormone-treated, unpollinated flowers was greater when the cluster was thinned to a single flower than in four-flowered clusters.

It can be concluded, therefore, that between-flower competition in Cox apple can reduce initial set markedly in some, but not all, years; and that, when competition is strong, pollinated flowers are more able to compete. This suggests that the seasonal status of some substrate such as carbohydrate might be limiting to initial set since the intensity of competition varies from year to year. But the competitive strength of flowers with fertilized ovules also emphasizes the potential importance of hormonal effluxes.

What is the evidence for limited carbohydrate availability? In temperate perennial fruits, carbohydrate reserves accumulated the previous summer and autumn are stored overwinter and then re-mobilized in the following spring to support the early enlargement of the flower cluster. Carbohydrate is added to the available pool by photosynthesis in the rosette leaves, but it appears that these leaves only become net exporters in apple after the cluster has reached the pink-bud stage (Quinlan & Preston, 1971; Vemmos, 1990). Evidence from experiments in which the movement of labelled carbohydrate to the flowers from rosette leaves and stored reserves was studied (Goldwin, 1985*b*) questioned whether flower expansion was entirely dependent on these sources during initial set. Subsequent work (Goldwin, 1989*b*; Vemmos, 1990) has shown that photosynthesis in the sepals and receptacle of apple flowers makes an important contribution to the carbohydrate pool at this time. In other species, flower photosynthesis has also been recorded (e.g. citrus buds (Vu, Yelenosky & Bausher, 1985) and ovary in *Brassica napus* (Addo-Quaye, Scarisbrick & Daniels, 1986)); the existence of this third source of carbohydrate suggests that leaf photosynthesis might not be such a vital contributor to the carbohydrate pool during initial set as might be expected. Indeed, the removal of all the rosette leaves in apple clusters does not prevent initial fruit set and growth of some fruits through to maturity, although set can be reduced. For example, in 1986, Cox apple clusters were thinned to four flowers and all the rosette leaves were removed four days before anthesis. Groups of clusters with leaves retained were used as controls. The flowers were decapitated to prevent pollination when at the balloon stage, and hormone-treated the following day. Records of set taken 2, 3 and 4 weeks after anthesis (control, 88, 78 and 55%; and de-leafed, 88, 67 and 38%, respectively) showed that leaf removal had little effect on initial set, but differences became more apparent as 'June drop' approached. Subsequent work by Vemmos (1990) showed that if the rosette leaves were removed earlier, between green cluster and the early balloon stage, the initial parthenocarpic set was reduced significantly.

With pollinated set, removal of rosette leaves has given similar results. Thus leaf removal shortly before bloom on Golden Delicious had no effect on set when recorded about three weeks after full bloom, but set was reduced three weeks later (Ferree & Palmer, 1982). Earlier removal of leaves of Cox trees from the mouse-ear stage to the pink-bud stage (Arthey & Wilkinson, 1964; Llewelyn, 1963, 1968) caused much greater reductions in set, with the earliest treatments having the greatest effect despite a much smaller leaf area removed. Arthey & Wilkinson (1964) and Vemmos (1990) concluded that leaf-removal effects were more likely to be due to changes in hormone status than to photosynthate reduction, because leaf removal at a stage when the leaves were not net exporters of photosynthate was more effective than when they were.

Shading of Cox apple trees during the period from flowering to leaf fall, which would be expected to deplete the carbohydrate pool stored over winter, had little effect on the initial set of Cox apple the following year, except when about 90% of the light was excluded (Jackson & Palmer, 1977).

With apple, then, the balance of evidence suggests that the failure of some flowers to set initially is unlikely to be due to an inferior capacity to compete for carbohydrate since, normally, carbohydrate supply is unlikely to be limiting. There is little evidence regarding competition for other substrates, although it is perhaps relevant that additional application of nitrogenous fertilizer in summer or autumn has been found to enhance the ability of apple flowers to set fruit the following spring (Williams, 1965). As with carbohydrate, nitrogen is also stored over winter in perennial fruit species in the form of protein and re-mobilized in the spring as amino acids, particularly arginine. The storage and mobilization of nitrogenous compounds in fruit trees has been reviewed by Tromp (1970). Some support for the potential importance of nitrogen-containing compounds comes from the work of Buszard (1983) who found that flower buds on trees previously de-fruited contained a higher concentration of arginine in the autumn than did buds on heavily cropped trees. Because the initial set of flowers the following spring in the clusters on the de-fruited trees was higher than those on heavily cropped trees, it is possible that, in situations of greater amino-acid availability, competition for this substrate is reduced. Clearly, much more evidence is required before a link between an inability to compete for amino acid and set failure is established; it is possible that the improved ovule longevity of flowers on trees previously treated with nitrogen (Williams, 1965) or with a reduced crop (Williams, 1970) is the real explanation for observed differences in set.

Although the evidence suggests that competition can prevent the initial set of some flowers, this cannot be the complete explanation for set failure because, even when inflorescences are thinned to a single flower, a large proportion can still fail to re-establish growth in response to hormone applied at anthesis (e.g. Fig. 1a, 30%; in other experiments, this proportion is as high as 80%).

The mode of hormone action in stimulating parthenocarpy
In order to clarify what other internal and environmental factors result in initial set failure, it seems sensible to establish firstly where the hormone acts in stimulating parthenocarpic set. Apple spurs are a convenient model system because they consist essentially only of rosette leaves, flowers and peduncle during the time when initial set is determined. At this stage, vegetative growth is minimal because the 'bourse' shoot has not yet started to enlarge and hence constitutes a small sink. With normal spray application, the hormone treatment hits all components of the cluster. Several modes of action can therefore be envisaged. Hormone alighting on the rosette leaves might promote export of photosynthate to the flower; hormone alighting on the pedicel might prevent abscission at the pedicel base; or hormone alighting on the flower might enhance the metabolic activity of the receptacle–ovary. In an experiment reported earlier (Goldwin, 1989a) with decapitated Cox flowers thinned to two flowers per cluster, the 'Wye' hormone mixture was applied by syringe: only to the calyx 'well' of flowers; to the pedicel base only; or to all the rosette leaves in the cluster. The initial set (61%) for the calyx treatment was very much higher than for treatments to the pedicel base (9%) or rosette leaves (13%). When hormone mixture containing ^{14}C-labelled GA_3 (the essential component in the mixture) was applied to the pedicel base, little movement of radioactivity to the receptacle or calyx parts of the flower occurred (Goldwin, 1989a), whereas application of the mixture to the calyx well resulted in appreciable movement of label to the receptacle and pedicel. Thus treatment to the pedicel base, which moved very little to the receptacle, was not very effective, whereas treatment to the calyx, which did move to the receptacle, was highly effective. These results strongly suggested that the site of action of the hormone lay in the receptacle–ovary. However, it was just conceivable that the alternative hypothesis of hormone inhibition of abscission was still tenable. If lateral movement of pedicel-applied hormone to the abscission zone did not occur, then this treatment would not have prevented abscission and a low set would have resulted (as observed). In contrast, the clearly demonstrated basipetal transport of hormone from the calyx would have been expected to reach the cells in the abscission zone through the vascular system.

In a further experiment carried out by the author to investigate Cox flower abscission, flower clusters on Cox's Orange Pippin apple trees were thinned to two flowers at the late 'balloon' stage of development, the 'king' flower always being removed. For 40 clusters, the flowers were severed at the base of the receptacle to leave only the pedicels attached to the peduncle. Lanolin was smeared on the pedicel cut surface to prevent desiccation. For a further 40 clusters, the flowers were left intact except that the petals, anthers and style were removed ('decapitation') to prevent pollination. On the same day, half of the clusters with pedicels and half of those with flowers were sprayed with the 'Wye' hormone mixture, the remainder being sprayed with water to act as controls. The clusters were inspected at daily intervals to determine the date of abscission by using very slight pressure to ascertain whether the organ was just about to detach.

The results showed that the abscission of control pedicels was first recorded eight days after treatment and was complete three days later (Fig. 2). In contrast, pedicels that had been hormone-treated were first observed abscising 11 d after treatment. The rate of abscission for this treatment was slower than for the controls; abscission was not complete before 25 d after treatment. The dates of 50% abscission for the two treatments were approximately seven days apart. The abscission of decapitated flowers generally occurred later than for pedicels. Thus, unpollinated control flowers did not start dropping before the tenth day after treatment, with abscission complete 24 d after treatment (Fig. 3). Hormone-treated, unpollinated flowers segregated into two populations: those that set initially and those that did not. Abscission of unsuccessful flowers also started on the tenth day after treatment and finished 24 d after treatment. The date of 50% abscission of unpollinated, control flowers occurred about 14 d after treatment, whereas the date of 50% abscission of hormone-treated, unsuccessful flowers was about 17 d after treatment. The rapid enlargement of the receptacles of successful flowers was first observed eight days after treatment, well before abscission in most of the unsuccessful flowers occurred.

The results of this experiment showed that the hormone treatment did, indeed, inhibit the abscission of pedicels and, to a lesser extent, of unpollinated flowers that failed to set. However, the prevention of abscission was not the primary action of the treatment: the re-establishment of ovary enlargement in successful flowers started many days before most unsuccessful flowers dropped. This order in the timing of growth and abscission has also been observed in other years. The time that normally elapses between the conditions occurring which induce abscission and the organ actually dropping is thought to be a matter of hours rather than weeks

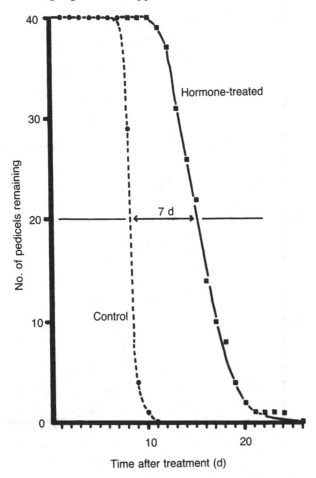

Fig. 2. Time courses of abscission of Cox flower pedicels treated either with water (control; circles) or with the 'Wye' hormone setting mixture (squares). There were 40 pedicels per treatment initially.

(Sexton & Roberts, 1982); it is therefore likely that the hormone induced rapid ovary growth well in advance of any abscission-delaying effects. It can be concluded that unsuccessful flowers do not set because the mechanism by which hormone re-establishes receptacle growth fails. This failure is not due to lack of hormone penetration or slow penetration, because hormone has been shown to move rapidly into the flower, with gibberellin being detected at the base of the pedicel by the time that an aqueous application to the calyx had dried (Goldwin, 1984).

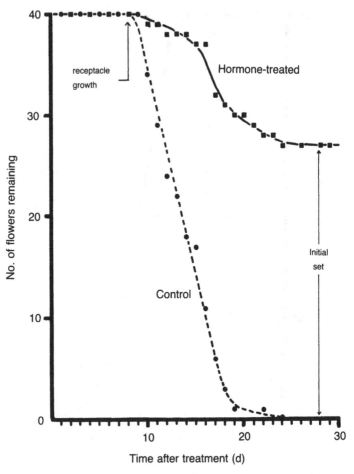

Fig. 3. Time courses of abscission of unpollinated Cox flowers treated either with water (control; circles) or with 'Wye' hormone setting mixture (squares). There were 40 flowers per treatment initially.

The effect of temperature on parthenocarpic set

Most studies of fruit setting have recorded only the numerical success of initial set by fruitlet counting after the successful and unsuccessful populations have separated (often after the unsuccessful population has abscised). This is not surprising: a detailed study of the growth of individual flowers is not only time-consuming but also difficult to achieve owing to the relatively small size of the flowers of many fruit species and the physical damage caused by the repeated use of vernier calipers. The popularity of cucurbits in studies of fruitlet development (e.g. Sedgley, Newbury & Possingham, 1977; Fuller & Leopold, 1977) and the fact that most cucurbits have a large ovary at anthesis are, perhaps, not unconnected. The alternative strategy of using time-lapse photography to study ovary growth is restricted (i) by the need to maintain the flower in a fixed position relative to the camera (not easy with flowers in an orchard) and (ii) by the unpredictability of set, which necessitates the use of many cameras to ensure the monitoring of setting and non-setting flowers with reasonable replication.

Faced with these problems, the author devised a photographic technique (Goldwin, 1985a) in which the silhouettes of the same flowers were obtained daily, and used to construct growth curves of receptacle diameter for many Cox flowers in the orchard. By using 100 unpollinated flowers, with different dates of anthesis and hormone treatment, it was possible to separate environmental and internal factors affecting set. Surprisingly, all flowers that responded to the treatment re-initiated rapid receptacle growth at about the same time. The experiment was repeated in the following year with the same result (Goldwin, 1985a). Comparison of data in the two years showed that the timing of growth coincided with an increase in temperature above 15 °C. Because, once started, growth did not cease when the temperature remained below 15 °C for several days in the second experiment, it was concluded that a temperature 'trigger' was required to initiate parthenocarpic growth in hormone-treated Cox flowers. This hypothesis has been confirmed in subsequent work (G.K. Goldwin & M. Herrero, unpublished data), in an experiment in which hormone-treated flowers were maintained artificially above 15 °C in the orchard during two successive nights at anthesis. During this period, the ambient temperature did not reach 15 °C. As anticipated, a proportion of these flowers were very soon 'triggered' into growth by the heating treatment, whereas flowers that were not heated only started growing several days later when the ambient temperature rose above 15 °C. However, a few heated flowers failed to grow earlier but were also triggered later by the rise in ambient temperature. This suggested that some flowers were immature at the time of heating and were only capable of responding to the hormone treatment later.

Parthenocarpic set in relation to ovule development

Under natural conditions of pollination, there is a period, which can last many days, between the time (petal opening) at which pollen can first be deposited on the stigma and the time at which fertilization takes place. The expenditure of plant resources on continued ovary growth of flowers before the success of fertilization has been determined is not advantageous. The observation that some apple flowers were incapable of enlarging just after anthesis, when hormone-treated and warmed, suggests that a mechanism for inhibiting ovary growth exists. A similar suggestion was made by Thompson (1961) who found that strawberry receptacle tissue failed to enlarge in the vicinity of unfertilized carpels, but did enlarge immediately below a fertilized one. Localized suppression of carpel formation enabled receptacle enlargement to occur. Thompson postulated that the effect of pollination or fertilization is to remove or suppress the growth inhibition.

In considering a possible mechanism for ovary growth inhibition, the most likely causative agents for control would be organs in which clearly defined developmental stages exist. Petal enlargement involves considerable utilization of resources (in Cox apple, the petals constitute about half the dry mass of the flower at anthesis) and it might be argued that ovary growth slows because available carbohydrate supply is diverted to support petal opening and nectar production. However, this explanation seems untenable because petal removal well before flower opening has little effect on the growth characteristics of the ovary. A more likely candidate for the controlling agent is the ovule, in which discrete, developmental changes occur (embryo-sac formation and maturation, and fusion of polar nuclei). Furthermore, the ovules are the most appropriate controlling agent, because it is their survival as viable seeds that constitutes the purpose of fruit production.

Investigation of the time courses of embryo sac development have been made in pear (Herrero & Gascon, 1987) and apple (G.K. Goldwin & M. Herrero, unpublished data); Costa Tura & MacKenzie (1990) have recorded the development of the apple ovule at different stages from 'mouse-ear' to petal fall. Goldwin & Herrero (1990) found that embryo sac maturation was complete several days before the time at which setting, parthenocarpic apple and pear flowers first started to grow rapidly. The onset of rapid growth coincided with the time at which about 50% of the polar nuclei had fused. It is possible that ovaries are released from inhibition when fusion of polar nuclei occurs.

In unpollinated flowers that receive no setting stimulus, the ovules eventually degenerate. In apple and pear, this occurs several days before flower abscission. Ovule longevity varies from year to year and can be affected by cultural conditions. Thus, nitrogenous fertilizer application to apple trees in summer or autumn resulted in an increase in ovule longevity and in the effective pollination period (EPP) (Williams, 1965). High temperatures during flowering shorten ovule longevity but, since the rate of pollen-tube growth is increased, this does not necessarily result in a lower probability of set. In the same way that pollination has been found to delay ovule degeneration by stimulating ovary growth, the ovules in hormone-treated flowers remain turgid in growing fruitlets. This effect has been observed (Goldwin, 1985*a*) several weeks after anthesis. Nucellus tissue in growing, parthenocarpic Cox fruitlets was apparently still viable up to 21 d after rapid growth first started, although the embryo sacs had degenerated by then. In contrast, fruitlets, which were removed at the same time but had stopped growing, contained only degenerated ovules, as did samples of growing fruitlets taken after 21 d.

The association between ovule turgidity and active receptable growth, which has been observed over many years in parthenocarpic pome fruits, suggested that receptacle growth might be dependent on ovule viability in the period critical for initial set. Other circumstantial evidence relating to apple receptacle growth *in vitro*, an association between ovule size and fruitlet size, and the effects of ovule destruction on fruitlet survival, tended to support this hypothesis (Goldwin, 1985*a*). Earlier, Sedgley, Newbury & Possingham (1977) and Sedgley (1979) postulated that the nucellus and integuments of unfertilized ovules promoted the development of auxin-induced, parthenocarpic water melon fruit, although Weinbaum & Simons (1974) and, later, Sedgley (1981) concluded that embryo abortion in pollinated fruits was not the cause of abscission in apple and *Macadamia*, respectively. Recent work by G.K. Goldwin & M. Herrero (unpublished data) has now cast doubt on the hypothesis that early fruitlet growth requires the presence of viable ovule tissue since frost-damaged pear flowers have been observed to set parthenocarpically, even though the ovules were destroyed at anthesis. Furthermore, recent, as yet unpublished, work by the author has shown that the growth *in vitro* of apple receptacle tissue from unpollinated flowers and young fruitlets is not promoted by the presence of viable ovules. However, this, in itself, is not proof of a lack of a promotory role for ovules because conditions *in vitro* and *in vivo* differ widely. For example, in the hormone-free medium used, growth occurred almost exclusively as a result of cell enlargement, whereas, on the tree, cell division is a major contributor to growth immediately after initial set (cell numbers in apple increasing from about 2 million at anthesis to about 20 million three weeks later) (Pearson & Robertson, 1953).

In summary, the balance of evidence suggests that, in pome fruits, the onset of rapid enlargement of the receptacle which signifies initial parthenocarpic set, prevents the degeneration of the nucellus and integuments rather than that growth is dependent on viable ovule tissue. In flowers that fail to respond to the setting stimulus, it is suggested that ovule degeneration is the first marker of an ensuing sequence of events in which the final act is flower abscission.

Environmental and internal factors regulating initial set

What, then, are the environmental and internal factors that regulate initial set? If apple and pear are taken as the model, evidence suggests that a 'window' for setting exists, during which time certain conditions must be fulfilled. The 'opening' of the 'window' occurs as the flower reaches a state of maturity which, it is suggested, coincides with polar nuclei fusion in the embryo sacs. Before this stage, it appears that ovary growth is inhibited by some mechanism related to the early development of the embryo sacs. 'Closing' of the 'window' is signalled by the onset of ovule degeneration. For parthenocarpy, hormone treatment must be applied before the 'window is closed', but also the temperature must exceed about 15 °C at some stage after the treatment has been applied and before the 'window is closed'. In England, favourable weather does not always occur at the required time, and it is likely that much of the variability in response to hormone treatment (Goldwin, 1978) has resulted for this reason.

When pollination is the stimulus for set, favourable weather must firstly allow pollen transfer, and then temperature must be high enough to ensure an adequate rate of pollen-tube growth down the style to a position where ovary growth is stimulated. By this time the 'window' has probably already 'opened'. If ovary growth is stimulated, then embryo-sac degeneration is delayed, allowing fertilization eventually to take place. This model for pollinated set differs slightly from that expounded by Williams (1970) who did not allow for the synergistic effects of receptacle growth in his definition of the 'effective pollination period' (EPP) as 'the time from flower opening to degeneration of the egg apparatus less the time taken for pollen tube growth from stigma to embryo sac'. In the situation where ovary growth is stimulated, the EPP will be increased.

Regulation of fruiting by effects on fruitlet survival

The factors that determine whether a flower survives the transition to a young fruitlet have been described at length, but a further major obstacle has to be overcome before fruit maturity is finally reached. With many species, a significant proportion of enlarging fruitlets abscise, usually during well-defined periods of growth. With tree fruits, the pronounced wave of fruitlet abscission that usually occurs several weeks after the initial set has taken place is termed 'June drop', even though this does not occur in June in many countries or with all species. The cause of this abscission has often been attributed to competition between fruitlets.

Evidence to support the hypothesis that competition is for carbohydrate has come from shading and leaf-removal experiments, while an hormonal mechanism has often been proposed as the way in which sinks compete. The presence of developing seeds in fruit has been shown to confer a competitive advantage on fruits, which has often been attributed to a greater hormonal efflux from these fruits or to an enhanced sink strength of seeded fruits owing to a larger hormone content.

Competition between fruitlets

The experiment described earlier, in which different numbers of parthenocarpic fruitlets were set in Cox clusters, shows that competition exists between fruitlets even in the absence of fertilized seeds, since clusters initially setting more fruitlets lost proportionately more during June drop (Fig. 1*b*). Indeed, in this experiment, there appeared to be an upper limit for final set of just over one fruit per cluster on average. In previous experiments (Goldwin, 1984), seeded and parthenocarpic fruitlets were set in the same cluster, but, despite starting to grow earlier and hence constituting larger sinks, the parthenocarpic fruitlets dropped to leave only seeded fruit remaining after 'June drop'. The period during which the presence of seeds is important for the survival of apple fruitlets was shown by Abbott (1959) to be relatively short. Seed removal from fruits after 'June drop' had finished had no effect on abscission, whereas prior to that, fruits all abscised.

Evidence that carbohydrate pool size limits fruitlet survival

Earlier in this paper, the concept that initial fruit set might be limited under normal conditions by the availability of carbohydrate was questioned. The evidence for subsequent fruitlet retention being affected by environmental effects on photosynthate production appears stronger, but not conclusive.

There have been several reports of leaf removal increasing 'June drop'. Thus, Llewelyn (1968) found that spur leaf removal on Lord Lambourne apple trees four weeks after full bloom caused a significant increase in fruit drop two weeks later, whereas leaf removal from bourse shoots had less effect. Quinlan & Preston (1971) found with three apple cultivars that greater proportions of fruitlets were lost during 'June drop' if bourse shoots were prevented from growing by repeated removal. This effect was shown to be due to the absence of bourse shoot leaves: the fruitlets still dropped if the shoots were allowed to remain, but without leaves.

Rosette leaf removal from Golden Delicious (Ferree & Palmer, 1982) and Cox (see earlier in this chapter) apple clusters shortly before bloom also increased 'June drop'.

There have been several reports on the effects of shading on fruit retention. Meadley & Milbourn (1971) found that shading of pea plants to half the light intensity of full daylight increased pod abscission more when the treatment was applied after flowering had started than when carried out before flowering. Jackson & Palmer (1977) shaded Cox apple trees so as to receive 37, 25 or 11% of full daylight during the period from flowering to leaf fall. Fruitlet retention was reduced by all three levels of shading, although, under even the most intense shading, retention did not drop below 40% of control levels. Byers *et al.* (1985) shaded peach and apple trees for ten-day periods to give 10% light on an overcast day. The effects of shading on fruit retention depended very much on the timing of the treatment. With peach, 10 d shading starting 15, 20, 25 or 45 d after full bloom had little effect on final fruit numbers, whereas shading starting 30, 35 or 40 d after full bloom greatly reduced the crop. The importance of the timing was also striking when apple branches were similarly treated, with only shading starting 16 d after full bloom having a highly significant effect on fruitlet retention. Earlier or later timings had little effect.

In this work, application of terbicil, an inhibitor of photosynthesis, also increased fruit drop; the authors suggested that reduced levels of particular photosynthates might limit fruit and/or seed development, thus triggering processes leading to fruit abscission. However, several aspects of this work are puzzling. Firstly, the effects of terbicil on the fruit retention of peach and apple did not follow exactly the same trend as shading, which would be expected if a reduction in the carbohydrate pool were the cause of 'June drop'. Secondly, it would be surprising if carbohydrate were only limiting in apple two to three weeks after full bloom, at a time when the rosette leaves were fully expanded and when the total content of sugars and sorbitol in spur and two-year wood is increasing (Vemmos, 1990). It seems therefore that, although leaf removal and shading undoubtedly reduce the carbohydrate pool available to support fruitlet expansion, under normal conditions it is questionable whether supplies are limited enough to trigger fruitlet abscission.

It is possible that other forms of stress afflicting the whole plant might be the true cause of fruitlet abscission, as it has been observed that periods when the growth rates of fruits and shoots decline are associated with the abscission of both leaves and the slowest-growing fruitlets (Zucconi, 1981). Low temperature and limited water availability, which also limit growth, are potential causes of fruit abscission.

Conclusions

There are still large gaps in our knowledge of the mechanisms controlling fruiting, particularly those related to ovary growth during the crucial period from just before petal opening to petal fall. In recent years, very detailed and elegant work has enabled the sequence of events from pollen release to pollen-tube growth to be elucidated at the biochemical and cellular level. In contrast, little appears to be known about the stimulus that induces the burst of cell division in the ovary indicative of successful initial set. Does this signal originate in the pollen tube or in the maternal cells through which the tube grows? Is the signal hormonal in nature (as perhaps the success of hormone-induced parthenocarpy in some species might suggest) and how does it interact with effluxes from the ovules, which appear to be inhibitory to growth at anthesis? What effect does fertilization have on effluxes from the ovules? The answers to these questions will only come through investigations carried out with the same detail that has been applied to studies of the pollen–style interaction.

Further studies on the subsequent abscission of apparently set fruitlets are also required, because it is far from clear why this occurs. It has been shown that extreme treatments such as heavy shading or leaf removal can induce fruit drop, yet the evidence is by no means conclusive that competition for a restricted carbohydrate pool is the stimulus for fruit abscission under normal conditions.

Fruits form an important part of the diet of the human population. It is surprising that so little is still known about the mechanisms that determine whether a flower develops to become a mature fruit. An intensified research effort in this field is surely long overdue.

Acknowledgements

The author is indebted to the Ministry of Agriculture, Fisheries and Food for financial support. Mr R. Smith is thanked for technical assistance.

References

Abbott, D.L. (1959). The effects of seed removal on the growth of apple fruitlets. *Report of Long Ashton Research Station for 1958*, pp. 52–6.

Abbott, D.L. (1973). The effect of autumn temperature on flower initiation and fruit bud development. *Report of Long Ashton Research Station for 1972*, pp. 34–5.

Abbott, D.L. (1977). Fruit-bud formation in Cox's Orange Pippin. *Report of Long Ashton Research Station for 1976*, pp. 167–76.

Addo-Quaye, A.A., Scarisbrick, D.H. & Daniels, R.W. (1986).

Assimilation and distribution of ^{14}C photosynthate in oilseed rape (*Brassica napus* L.). *Field Crops Research* **13**, 205–15.

Arthey, V.D. & Wilkinson, E.H. (1964). The effect of pre-blossom defoliation on the cropping of Cox's Orange Pippin apple. *Horticultural Research* **4**, 22–6.

Browning, G. (1989). The physiology of fruit set. In *Manipulation of Fruiting* (ed. C.J. Wright), pp. 195–217. London: Butterworths.

Buszard, D.J.I. (1983). *Flower quality in relation to apple fruit set promotion by hormone sprays*. Ph.D. thesis, University of London.

Buszard, D.J.I. & Schwabe, W.W. (1989). Effect of crop history on response of apple cv Cox's Orange Pippin to a range of fruit setting techniques. *Acta Horticulturae* **239**, 353–6.

Byers, R.E., Lyons, C.G. Jr, Yoder, K.S., Barden, J.A. & Young, R.W. (1985). Peach and apple thinning by shading and photosynthetic inhibition. *Journal of Horticultural Science* **60**, 465–72.

Chapman, G.P., Fagg, C.W. & Peat, W.E. (1979). Parthenocarpy and internal competition in *Vicia faba* L. *Zeitschrift für Pflanzenphysiologie* **94**, 247–55.

Coombe, B.G. (1976). The development of fleshy fruits. *Annual Review of Plant Physiology* **27**, 507–28.

Costa Tura, J. & MacKenzie, K.A.D. (1990). Ovule and embryo sac development in *Malus pumila* L. cv Cox's Orange Pippin, from dormancy to blossom. *Annals of Botany* **66**, 443–50.

Crane, J.C. (1964). Growth substances in fruit setting and development. *Annual Review of Plant Physiology* **15**, 303–26.

El-Nadi, A.H. (1969). Water relations of beans. I. Effects of water stress on growth and flowering. *Experimental Agriculture* **5**, 195–207.

Ferree, D.C. & Palmer, J.W. (1982). Effect of spur defoliation and ringing during bloom on fruiting, fruit mineral level, and net photosynthesis of 'Golden Delicious' apple. *Journal of the American Society for Horticultural Science* **107**, 1182–6.

Free, J.B. (1970). *Insect Pollination of Crops*. London: Academic Press.

Fuller, G.L. & Leopold, A.C. (1976). Pollination and the timing of fruit-set in cucumbers. *HortScience* **10**, 617–19.

Fuller, G.L. & Leopold, A.C. (1977). The role of nucleic acid synthesis in cucumber fruit set. *Journal of the American Society of Horticultural Science* **102**, 384–8.

Goldwin, G.K. (1978). Improved fruit setting with plant growth hormones. *Acta Horticulturae* **80**, 115–21.

Goldwin, G.K. (1981). Hormone-induced setting of Cox apple, *Malus pumila*, as affected by time of application and flower type. *Journal of Horticultural Science* **56**, 345–52.

Goldwin, G.K. (1984). Factors affecting hormone-assisted setting of Cox's apple. *Acta Horticulturae* **149**, 161–71.

Goldwin, G.K. (1985a). The role of ovules in the survival of hormone-induced parthenocarpic apple flowers. In *Experimental Manipulation of Ovule Tissues* (ed. G.P. Chapman, S.H. Mantell & R.W. Daniels), pp. 89–108. London: Longman.

Goldwin, G.K. (1985*b*). The use of plant growth regulators to improve fruit setting. In *Growth Regulators in Horticulture* (ed. R. Menhenett & M.B. Jackson), pp. 71–88. Long Ashton, England: British Plant Growth Regulator Group.

Goldwin, G.K. (1986). Use of hormone setting sprays with monoculture orchards to give more regular cropping. *Acta Horticulturae* **179**, 343–8.

Goldwin, G.K. (1989*a*). Improved fruit set in apple using plant hormones. In *Manipulation of Fruiting* (ed. C.J. Wright), pp. 219–32. London: Butterworths.

Goldwin, G.K. (1989*b*). Hormone-induced parthenocarpy as a system for studying flower 'quality' in Cox's Orange Pippin apple. *Acta Horticulturae* **239**, 349–52.

Goodwin, P.B. (1978). Phytohormones and fruit growth. In *Phytohormones and Related Compounds – A Comprehensive Treatise*, volume II (ed. D.S. Letham, P.B. Goodwin & T.J.V. Higgins), pp. 175–214. Amsterdam: Elsevier–North Holland.

Gustafson, F.G. (1937). Parthenocarpy induced by pollen extracts. *American Journal of Botany* **24**, 102–7.

Herrero, M. & Gascon, M. (1987). Prolongation of embryo sac viability in pear (*Pyrus communis*) following pollination or treatment with gibberellic acid. *Annals of Botany* **60**, 287–93.

Hoad, G.V. (1978). The role of seed derived hormones in the control of flowering in apple. *Acta Horticulturae* **80**, 93–104.

Howlett, F.S. (1927). Some factors of importance in fruit setting studies with apple varieties. *Proceedings of the American Society of Horticultural Science for 1926*, pp. 307–15.

Jackson, J.E. & Palmer, J.W. (1977). Effects of shade on the growth and cropping of apple trees. II. Effects on components of yield. *Journal of Horticultural Science* **52**, 253–66.

Keulemans, J. & van Laer, J. (1989). Effective pollination period of plums: The influence of temperature on pollen germination and pollen tube growth. In *Manipulation of Fruiting* (ed. C.J. Wright), pp. 159–71. London: Butterworths.

Kinet, J.M., Hurdebise, D., Parmentier, A. & Stainier, R. (1978). Promotion of inflorescence development by growth substance treatments to tomato plants grown in insufficient light conditions. *Journal of the American Society for Horticultural Science* **103**, 724–9.

Lever, B.G. (1985). The economics of PGR research – risk or reward. In *Growth Regulators in Horticulture* (ed. R. Menhenett & M.B. Jackson), pp. 15–28. Long Ashton, England: British Plant Growth Regulator Group.

Lewis, D. (1942). Parthenocarpy induced by frost in pears. *Journal of Pomology* **20**, 40–1.

Llewelyn, F.W.M. (1963). The importance of spur leaves and lime-sulphur sprays on fruit retention in three apple varieties. *Report of East Malling Research Station for 1962*, pp. 89–92.

Llewelyn, F.W.M. (1968). The effect of partial defoliation at different times in the season on fruit drop and shoot growth in Lord Lambourne apple trees. *Journal of Horticultural Science* **43**, 519–26.

Meadley, J.T. & Milbourn, G.M. (1971). The growth of vining peas. III. The effect of shading on abscission of flowers and pods. *Journal of Agricultural Science* **77**, 103–8.

Modlibowska, I. (1945). Pollen tube growth and embryo-sac development in apples and pears. *Journal of Pomology* **21**, 57–89.

Modlibowska, I. (1963). Effect of gibberellic acid on fruit development of frost damaged conference pears. *Report of East Malling Research Staton for 1962*, pp. 64–7.

Modlibowska, I. (1964). Frost damage and recovery in plant tissues. *Proceedings of the 16th International Horticultural Congress, Brussels 1962* **5**, 180–9.

Monselise, S.P. (1986). *CRC Handbook of Fruit Set and Development*. Boca Raton, Florida: CRC Press.

Moore, T.C. & Ecklund, P.R. (1975). Role of gibberellins in the development of fruits and seeds. In *Gibberellins and Plant Growth* (ed. H.N. Krishnasmorthy), pp. 145–82. New Delhi: Wiley Eastern.

Naylor, A.W. (1984). Functions of hormones at the organ level of organisation. In *Encyclopedia of Plant Physiology* new series, volume 10 (ed. A. Pirson & M.H. Zimmermann), pp. 172–218. Berlin: Springer-Verlag.

Nitsch, J.P. (1970). Hormonal factors in growth and development. In *The Biochemistry of Fruits and their Products*, volume 1 (ed. A.C. Hulme), pp. 427–72. London: Academic Press.

Nitsch, J.P. (1971). Perennation through seeds and other structures: fruit development. In *Plant Physiology: A Treatise*, volume VI A (ed. F.C. Steward), pp. 413–501. New York: Academic Press.

Pearson, J.A. & Robertson, R.N. (1953). The physiology of growth in apple fruits. IV. Seasonal variation in cell size, nitrogen metabolism and respiration in developing Granny Smith apple fruits. *Australian Journal of Biological Sciences* **6**, 1–20.

Quinlan, J.D. & Preston, A.P. (1971). The influence of shoot competition on fruit retention and cropping of apple trees. *Journal of Horticultural Science* **46**, 525–34.

Schwabe, W.W. & Mills, J.J. (1981). Hormones and parthenocarpic fruit set – A literature survey. *Horticultural Abstracts* **51**, 661–98.

Sedgley, M. (1979). Ovule and seed growth in pollinated and auxin-induced parthenocarpic watermelon fruits. *Annals of Botany* **43**, 135–40.

Sedgley, M. (1981). Early development of the *Macadamia* ovary. *Australian Journal of Botany* **29**, 185–93.

Sedgley, M., Newbury, H.J. & Possingham, J.V. (1977). Early fruit development in the watermelon: Anatomical comparison of pollinated, auxin-induced parthenocarpic and unpollinated fruits. *Annals of Botany* **41**, 1345–55.

Sexton, R. & Roberts, J.A. (1982). Cell biology of abscission. *Annual Review of Plant Physiology* **33**, 133–62.

Silva, J.M., Holgate, M.E. & Abbott, D.L. (1980). Some aspects of crop variation in 'Golden Delicious' apple trees. *Scientia Horticulturae* **13**, 27–32.

Smith, M.L. (1982). Response of four genotypes of spring faba beans (*Vicia faba* L. minor) to irrigation during the flowering period in the United Kingdom. *FABIS* **4**, 39–41.

Stephenson, A.G. (1981). Flower and fruit abortion: proximate causes and ultimate functions. *Annual Review of Ecology and Systematics* **12**, 253–79.

Thompson, P.A. (1961). Evidence for a factor which prevents the development of parthenocarpic fruits in the strawberry. *Journal of Experimental Botany* **12**, 199–206.

Treharne, K.J., Quinlan, J.D., Knight, J.N. & Ward, D.A. (1985). Hormonal regulation of fruit development in apple: 'A mini-review'. *Plant Growth Regulation* **3**, 125–32.

Tromp, J. (1970). Storage and mobilization of nitrogenous compounds in apple trees with special reference to arginine. In *Physiology of Tree Crops* (ed. L.C. Luckwill & C.V. Cutting), pp. 143–59. London: Academic Press.

Tromp, J. (1982). Flower-bud formation in apple as affected by various gibberellins. *Journal of Horticultural Science* **57**, 277–82.

Van Overbeek, J., Conklin, M.E. & Blakeslee, A.F. (1941). Chemical stimulation of ovule development and its possible relation to parthenogenesis. *American Journal of Botany* **28**, 647–56.

Vemmos, S. (1990). Photosynthesis and carbohydrate content of Cox's Orange Pippin apple flowers in relation to fruit setting. Ph.D. thesis, University of London.

Visser, T. & Oost, E.H. (1982). Pollen and pollination experiments. V. An empirical basis for a mentor pollen effect observed on the growth of incompatible pollen tubes in pear. *Euphytica* **31**, 305–12.

Vu, J.C.V., Yelenosky, G. & Bausher, M.G. (1985). Photosynthetic activity in the flower buds of 'Valencia' orange (*Citrus sinensis* L. Osbeck). *Plant Physiology* **78**, 420–3.

Weinbaum, S.A. & Simons, R.K. (1974). An ultrastructural evaluation of the relationship of embryo/endosperm abortion to apple fruit abscission during the post-blossom period. *Journal of the American Society for Horticultural Science* **99**, 311–14.

Williams, R.R. (1965). The effect of summer nitrogen applications on the quality of apple blossom. *Journal of Horticultural Science* **40**, 31–41.

Williams, R.R. (1970). Factors affecting pollination in fruit trees. In *Physiology of Tree Crops* (ed. L.C. Luckwill & C.V. Cutting), pp. 193–207. London: Academic Press.

Williams, R.R., Arnold, G.M., Flook, V.A. & Jefferies, C.J. (1980). The effect of picking date on blossoming and fruit set in the following year for the apple cv Bramley's Seedling. *Journal of Horticultural Science* **55**, 359–62.

Yasuda, S., Inaba, T. & Takahashi, Y. (1935). Parthenocarpy caused by the stimulation of pollination in some plants of the cucurbitaceae. *Agriculture and Horticulture* **10**, 1385–90. (Cited by Gustafson (1937). In Japanese with English résumé.)

Zeller, O. (1960). Entwicklungsgeschichte der Blütenknospen und Fruchtanlagen an einjahrigen Langtrieben von Apfelbüschen. I. Entwicklungsverlauf und Entwicklungsmorphologie der Blüten am einjährigen Langtrieb. *Zeitschrift für Pflanzenzüchtung* **44**, 175–214.

Zucconi, F. (1981). Regulation of abscission in growing fruit. *Acta Horticulturae* **120**, 89–94.

L.C. HO

Fruit growth and sink strength

Introduction

Fruit growth is part of the integrated growth of a plant. Therefore, fruit yield is determined by the interaction between growing conditions and morphological characters, as well as the physiological activities, of the whole plant.

It has long been recognized that the improvement of fruit yield is dependent on our understanding of the factors controlling both the production by the leaf (i.e. assimilate production) and the sink strength of the fruit (i.e. assimilate partitioning) (see Watson, 1968). Most likely, the key to understanding the regulation of fruit growth is to identify the responses to the environment by the morphological factors and metabolic processes inside the fruit interacting with those in the rest of the plant. In terms of assimilate partitioning, fruits are irreversible storage sinks, as the imported assimilate is either used for growth or stored as reserves and no net export occurs during the life of the organ (Ho, 1988). Therefore, it is essential to know how the supply, or the competition for the supply, of assimilate by the individual fruit is regulated. In this review, I examine fruit growth in terms of source–sink interaction, sink competition and sink strength determination. Evidence will be presented that, apart from the supply of assimilate, both cell number (i.e. sink size) and some of the physiological activities (i.e. sink activity) within a fruit may determine its sink strength in attracting assimilate to sustain fruit growth. Background information on the physiology of fruit growth has already been comprehensively reviewed by Bollard (1970) and Coombe (1976).

Fruit growth and photosynthesis

In general, fruit yield in terms of dry matter production is related to assimilate supply and hence to the irradiation intercepted by the crop. This relation does not imply that any increase in fruit yield is entirely determined by the improvement of photosynthesis. Instead, fruit yield can be enhanced by changing dry matter partitioning alone in most crop plants under defined growing conditions (Gifford et al., 1984). The improvement of fruit yield can be achieved by manipulating the sink strength of the fruit (Gifford & Evans, 1981; Ho, 1988). However, the fruit yield is related to the concurrent photosynthetic rate of the plant only when fruit yield is below its potential.

Yield and solar irradiation

In temperate regions, crop yield is often related to the solar irradiation intercepted. For instance, under similar cultivation conditions, the annual yield of glasshouse tomato is found to be linearly related to the number of sunny days, a measure of the total irradiation during crop growth (Bewley, 1929). The fruit yield, therefore, reduces in proportion to the degree of shading (Cockshull, 1988). However, this relationship operates only for individual fruit when the irradiation is above the minimum requirement for flower development, otherwise fruit will not set (Kinet, 1977). Once the potential number and size of the fruit are fixed, fruit growth will respond to any extra assimilate supply resulting from further increased irradiation, but will not extend beyond its potential; a plateau of fruit yield is expected.

Rate of fruit growth and the rate of photosynthesis

In general, fruit growth is sustained principally by the supply of current photoassimilate. Thus, the rate of dry matter accumulation in fruits of the same potential size will relate to the rate of concurrent photosynthesis in the leaves. For instance, the contribution of assimilate fixed by the tomato fruit itself is less than 15% (Tanaka, Fujita & Kikuchi, 1974) and is mainly for organic acid and subsequently for protein biosynthesis rather than for sugar synthesis (Laval-Martin, Farineau & Diamond, 1977; Farineau & Laval-Martin, 1977). When the potential photosynthetic rate is higher than the potential sink demand, fruit growth may increase the actual photosynthetic rate (Hansen, 1969), although this is not always the case (Roper et al., 1988). On the other hand, when the potential sink demand is higher than the potential photosynthetic rate, competition between fruit will be intensified and result in fruit abscission.

For monocarpic crops (i.e. plants with a single reproductive phase) such as cereals, grain growth is sustained by both the concurrent photosynthetic rate of the leaves and the re-mobilization of reserves from the stem (Rawson & Evans, 1971). About 39% of the stem dry matter loss is used for grain growth in wheat (Austin *et al.*, 1977). This contribution from the stem can account for 70% of the grain weight in barley (Gallagher, Biscoe & Scott, 1975). On the other hand, the growth rate of fruit of polycarpic crops (i.e. plants with a multiple reproductive phase) such as tomato is closely related to the concurrent photosynthesis of their adjacent leaves. When the potential sink demand is higher than the concurrent photosynthetic rate, the fruit can obtain assimilate from the stem reserve (Ho, 1979).

Commonly, fruit size is increased by a higher leaf:fruit ratio (see Hansen, 1989). With less competition between fruit there is a greater supply of assimilate to the remaining fruit. However, the determination of potential yield is not necessarily related to the duration of leaf growth. Delaying leaf senescence may not increase fruit yield once the potential number and size of the fruit have been set.

Competition between vegetative and reproductive organs

Fruit, as the final growth stage of a reproductive organ, is commonly a strong sink for assimilate. It has long been observed that developing fruit can attract assimilate wherever available at the expense of vegetative growth (see Bollard, 1970). However, the flower generally has a lower priority than fruit in attracting assimilates. In fact, early flower development can be retarded by severe competition from vegetative growth (Kinet, 1977). Once fruits start to develop, both the direction and pathway of assimilate transport change in favour of fruit growth (Ho & Hewett, 1986). For instance, the priority of sink competition in tomato changes from the order roots > young leaves > inflorescence in flowering plants to that of fruit > young leaves > flowers > roots in fruiting plants (Ho, Grange & Shaw, 1989). Therefore, it is necessary to examine the competition for assimilate and other reproductive organs in terms of inhibition or dominance (Fig. 1).

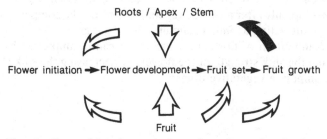

Fig. 1. Competition for assimilate throughout the development of reproductive organs.

The degree of competition for assimilates between vegetative and reproductive organs depends on their spatial and temporal relationships.

In annual plants, the initiation and development of reproductive organs occur in the same season. In monocarpic plants, fruit development takes place after the completion of the vegetative growth and may not be subjected to competition from the vegetative growth. Instead, the demand of assimilate for fruit growth is so predominant that the senescence of the whole plant starts at the completion of fruit growth. In heavily fruit-loaded apple trees, the dry matter accumulation in the roots falls from 40% in the non-fruiting tree to 10% of the total plant mass (Heim et al., 1979). Similar retarding effects of fruit to root growth have been observed in peach (Chalmers & Ende, 1975), tangerine (Smith, 1976) and tomato (Hurd, Gay & Mountifield, 1979). In general, the removal of fruit can improve vegetative growth early in plant development or delay the onset of leaf senescence at a later stage (Nooden & Guiamet, 1989). On the other hand, competition between the reproductive and vegetative organs in polycarpic plants, such as tomato, is very localized (one truss every 3–5 leaves) throughout the plant's development (Ho & Hewitt, 1986).

For perennial plants, the time span from initiation to the completion of the growth of reproductive organs can extend over several months. Floral buds may be initiated in the season preceding that in which fruit develop and floral initiation may be inhibited by concurrent fruit growth (Monselise & Goldsmidt, 1982). The development of fruit would depend on assimilates from the reserve in the tree stem as well as from the concurrently fixed assimilates in the leaves. The competition for assimilate in biennial plants such as raspberry, with vegetative shoots and reproductive shoots in alternate years, is quite different from that in most annual cropping plants such as apples. Two-thirds of the vegetative growth in the first-year shoot (i.e. primocane) is in the stem, but the growth of fruit in the second-year shoot (i.e. fruiting cane) is at the expense of the previous and current vegetative growth (Waister & Wright, 1986). As the flower buds develop only after a year of vegetative growth, the competition from flowers in attracting assimilate should not be too great.

The competition of flowers for assimilate can be improved by either increasing the sink strength of the flower, or decreasing the sink strength of the competing vegetative organs.

Increasing the sink strength of flowers in sink competition

Reproductive organs at flowering may or may not be a strong sink for assimilate. In tulip, the flower stem is the dominant sink for assimilate (Ho & Rees, 1977). In contrast, the tomato inflorescence is a very weak sink, which will abort readily when only a limited assimilate supply is available for young leaves and roots (Kinet, 1977; Russell & Morris, 1983). The competition against the flower can be reduced, not just by limiting the growth of competing organs by such means as root restriction (Cooper, 1964) or topping of the plant (Leopold & Lam, 1960), but by increasing the sink strength of the flower by hormonal treatment of the aborting inflorescence (Kinet *et al.*, 1978). The application of cytokinin and gibberellic acid (GA) to the aborting flower re-starts cell division in the ovary (Kinet *et al.*, 1986); this appears to increase the sink strength of the flower to divert assimilates from the apex (Leonard *et al.*, 1983). These researchers found that DNA synthesis increased in the ovules of the aborting flower less than 20 h after the hormonal treatment, followed by an increase of mitotic activity of the nuclei and an accumulation of carbohydrates in the inflorescence. Within 27 h, the import of assimilate to the inflorescence increased. The decrease of import into the apical shoot was detected between the first and third day after the treatment (Kinet, 1987) (Fig. 2).

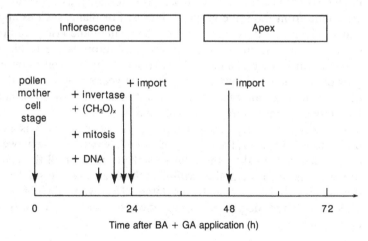

Fig. 2. The chronicle of events in the alteration of partitioning of assimilate after hormonal treatment to the aborting flower in tomato plants (after Kinet, 1987).

Reducing the sink strength of the competing organs in favour of flower in sink competition

In practice, fruit yield of tomato has been improved by applying growth retardants to reduce stem growth (Knavel, 1969). In nature, plants with dwarfing genes tend to have higher fruit yield. Through plant breeding, the higher yield of semi-dwarf wheat with Norin-10 dwarfing gene is coupled with a reduced growth of the stem (Brooking & Kirby, 1981). The shortening of the stem starts before floral initiation; consequently, the floral initiation period is prolonged and more grains are set (Fisher & Stockman, 1986). As floret formation occurs at the same time as cell division in the stem (Nilson, Johnson & Gardner, 1957), the genetic inhibition of cell division and cell elongation in the competing organs may be the cause of more florets. As the size of the grains is not increased, sink number rather than the sink strength of the individual sinks is enhanced. Similarly, yields of apple can be increased from 50% to 70% of harvest index by grafting onto dwarf rootstock (Barlow & Smith, 1971), which tend to have less vegetative shoot growth (Hansen, 1980).

Competition between fruit

As the fruit is a stronger sink than flowers and vegetative organs for assimilates, the competition for assimilates by a fruit is mainly with other, adjacent fruit. This competition is commonly observed when fruits grow in clusters (as in apple) or in trusses (as in tomato), as such fruits obtain assimilate from the same group of leaves. For plants such as apples, in which all fruits develop more or less simultaneously, competition between fruit can be very intense and fruit drop as the result of fruit competition is very common. Furthermore, competition between fruit growth and the floral initiation for the next year's crop may be the cause of the biennial or alternate bearing habit in a great number of perennial fruit crops (Monselise & Goldsmidt, 1982).

The mechanism regulating competition between fruits is not fully understood. However, the priority of assimilate partitioning between sink organs is determined by the relative sink strength of all the competing organs. This intrinsic ability within each organ may be related to cell division which is hormonally regulated. The availability of assimilate affects only the degree of competition but not the priority of the competition.

Dominance in fruit growth

The early-initiated fruit in some species inhibit the growth of the later-initiated ones. If the competition is between fruit and initiated flowers, flower initiation is inhibited (Buttrose & Sedgley, 1978). This dominance effect of early-set fruit over later-set fruit can either delay or limit the development of the latter. When assimilate is limited, fruit competition causes fruit drop, particularly among the later-set fruit, as frequently observed in apple (Abbot, 1984) or in beans (Tamas *et al.*, 1979). Quite often, the continuous growth of a dominant fruit inhibits not just the growth of other fruit, but the growth of the rest of the plant, resulting in early senescence (de Stigter, 1970).

The mechanism of dominance is not certain, but competition for assimilate between fruit may not be the sole cause as the amount of assimilate imported by the flowers or young fruit is quite small. As the dominant fruit in the competition is often the early-initiated one, the priority to obtain assimilates appears to be fixed by the fruit-set sequence. As shown in tomato, once the fruit-set sequence of the fruit in the same truss is synchronized, the difference in size between fruit can be greatly reduced (Bohner & Bangerth, 1988*a*). However, the dominance of early-set fruit of zucchini (*Cucurbita pepo*) is closely related to seed number. Apparently, the ovule number decreases with the sequence of flower initiation within an inflorescence in a number of species (Thomson, 1989). Indeed, once the seed number in the early-set fruit of zucchini was limited, more fruit were set at shorter time intervals. When seed number was manipulated, fruits with fewer seeds, regardless of fruit-set sequence, were inhibited by fruits with more seeds (Stephenson, Devlin & Horton, 1988). It appears that the dominance of fruit may be mediated by the amount of IAA produced by seeds of fruit on the same vine. It has been postulated that the efflux of IAA from the early-set fruit would cause an autoinhibition of the efflux of IAA from the later-set fruit at the junction where the IAA effluxes from different fruits meet. Assuming that the import of assimilate is regulated by the efflux of IAA from the individual fruit, less assimilate will be attracted by the later-set fruits, resulting in dominance of the early-set ones (Bangerth, 1989). This hypothesis suggests that a hormonal signalling mechanism is the basis for correlative dominance between fruits, resulting in fruit competition for assimilates.

Even in fruit without strong dominance characteristics, the fruits still compete with each other for assimilates if they have different sink strengths. In tomato, the early-set proximal fruit is potentially bigger than the later-set distal fruit of the same truss (Beadle, 1937), even when the sink competition is eliminated and the assimilate supply is abundant (Ho,

1980). This intrinsic difference between proximal and distal fruit may be mainly due to differences in cell number in the pre-anthesis ovary (Bohner & Bangerth, 1988a) and cannot be completely eliminated even when the fruit-set sequence is altered (Bangerth & Ho, 1984). In general, fruit can grow to its potential size only when fruit number is reduced to ensure an unlimited supply of assimilate.

Competition and compensation in fruit growth

For individual fruit, there are always a number of leaves acting as principal assimilate suppliers (Hansen, 1969; Khan & Sagar, 1966) in accordance with phyllotaxis (Russell & Morris, 1983). However, the number of leaves and their distance from individual fruit can be increased when fruit is thinned. When fruit number is reduced, assimilate targeted for the removed fruit will be attracted by the adjacent remaining fruit. As a result, the adjacent truss yield increases and the crop yield per plant will be compensated very closely to that of the control plants (Slack & Calvert, 1977).

Seed number and fruit size

It is generally accepted that the sink strength of a fruit is regulated by endogenous hormones related to cell division or seed development (Coombe, 1965; Crane, 1965). However, these two processes are important, but play different roles in regulating fruit growth.

Frequently, there are positive relationships between seed numbers in the fruit and fruit size. However, these relationships in tomato are variable with temperature and truss positions of the plant, and between genotypes (Imanishi & Hiura, 1975; Rylski, 1979). There is no constant ratio between fruit mass and seed number. In the same plant, fruit with seeds accumulate more dry matter and compete better for assimilate than fruit with fewer or no seeds. As a result, parthenocarpic fruit develops poorly in the presence of seeded fruit (Goldwin, 1984). However, the difference in fruit mass between seeded and the artificially induced parthenocarpic tomato fruit can be accounted for by the presence or absence of the seeds (Bangerth, 1981). Therefore, seed is not the predominant factor in determining the sink strength of a fruit. As the diffusible IAA from the fruit is related to the seed number and IAA may stimulate cell division or cell enlargement of the tissue around the seeds (Varga & Bruinsma, 1976), the role of seeds in producing IAA on fruit development may be important. However, early fruit development is mainly due to cell division before the peak of IAA production by the seeds. In kiwi

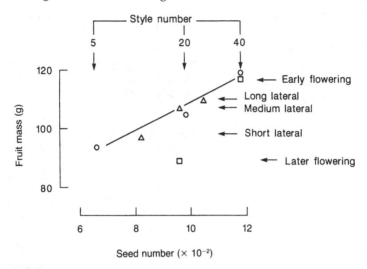

Fig. 3. The relationship between seed number and flesh mass of kiwi fruit (after Lai, Wolley & Lewes, 1990).

fruit, fruit size is linearly related to the seed number only in fruits with different style numbers or grown on different types of laterals (Lai, Wolley & Lewes, 1990) (Fig. 3). However, it is likely that the disproportionately smaller fruit size of the late-flowering fruit is due to a lower cell number in the pre-anthesis ovary as well as fewer seeds.

Cell number and potential fruit size

Although the mechanism regulating competition between fruits for assimilates is far from certain, it is most likely that competition is due to differences in sink strength between fruits. If sink size is the physical constraint and sink activity is the physiological constraint of sink strength, they should be quantified by parameters signifying the underlying mechanisms (Ho, 1988). At present, the cell number of the storage tissue of a fruit is considered a good measure of potential sink size, while the rates of limiting physiological processes for the uptake and accumulation of imported assimilate in the storage cells may be a meaningful measure of the sink activity (Fig. 4). In general, fruit mass is determined by cell number, cell size and cell content. The final volume of fruit, particularly those fleshy fruit that accumulate a lot of water, is mainly determined by the extent of cell enlargement (Coombe, 1976). While the actual fruit size is mainly affected by cell enlargement, the potential size of the fruit may

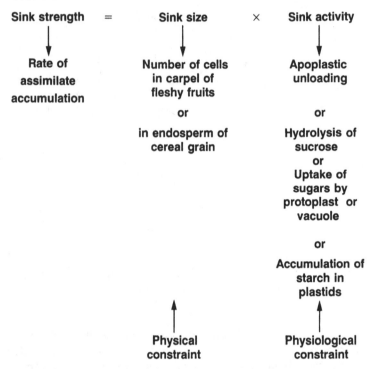

Fig. 4. A proposal to measure the sink size and sink activity to quantify the sink strength of fruit growth.

be determined by the final number of cells in the fruit. As observed in apricots, the size difference in fruit on the same tree was mainly due to differences in cell number rather than cell size (Jackson & Coombe, 1966). The duration of the cell division after anthesis varies from less than two weeks in tomato to more than four weeks in a number of stone fruits or even throughout fruit development as in avocado (Bollard, 1970). In fact, the duration of cell division varies between tissues of the same fruit. For instance in strawberry, cell division ceases soon after anthesis in the cortex but continues in the pith throughout fruit development (Havis, 1943). There is a tendency in fruits such as tomato and apricot, with a relatively short cell division period after anthesis, to have big cells with a high degree of polyploidy (Bradley & Crane, 1955; Bunger-Kibler & Bangerth, 1982). The rate of cell enlargement increases with the degree of ploidy. In comparison, fruit such as avocado with continuous cell division tend to have small final cell size but a big proportional fruit volume change of up to 300 000-fold from anthesis to maturity (Coombe, 1976). Considering that the duration of cell division after anthesis varies greatly, it is important to assess the relative importance of cell-division activity, before and after anthesis, in determining the final cell number.

Cell division before anthesis

Cell division in the ovary before anthesis may be important for fruit with a short period of cell division after anthesis. For instance, the cell generations (i.e. doublings) were about 20 in the ovary before anthesis but were only 2.5 – 4.5 after anthesis, over a period of less than four weeks in both apple (Pearson & Robertson, 1953) and tomato (Davies & Cocking, 1965; Bunger-Kibler & Bangerth, 1982; Teitel *et al.*, 1985; Bohner & Bangerth, 1988*a*). There is evidence that the difference in final fruit size in tomato at different positions on the truss (Bohner & Bangerth, 1988*a*) or in fruit of different genetic backgrounds (Bohner & Bangerth, 1988*b*) is essentially determined by the cell number in the pre-anthesis ovary. The potential difference in fruit size between cultivars of tomato can be assessed by the cell number in the pericarp of their pre-anthesis ovaries. The doublings in cell number to its maximum after anthesis in both proximal and distal fruit of the same cultivar are essentially the same (Fig. 5). Similarly, when fruit size is increased by irradiation mutation, the

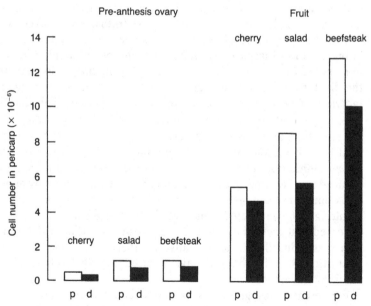

Fig. 5. The cell number of pre-anthesis ovary and fruit (14 d after anthesis) of different types of tomato at proximal (p) or distal (d) positions of the truss (L.C. Ho, unpublished).

increase is entirely accounted for by the increase of cell number in the pre-anthesis ovary as the subsequent cell division activity after anthesis is the same as the parent (Bohner & Bangerth, 1988b). The critical time and the effective means of altering the rate of cell division before anthesis are not known. However, the locule number of a tomato fruit can also be increased by applying GA to the apex before floral initiation, resulting in bigger fruit (Sawhney & Dabbs, 1978). It appears that the cell number of the pre-anthesis ovary can be manipulated throughout floral development.

Factors affecting final cell number

For a great number of plants, cell division in the ovary ceases just before anthesis, and the re-start of the cell division, whether due to fertilization or hormonal treatment, is the cause of increasing the sink strength of the fruit to attract assimilates. The immediate response has been demonstrated by the import of ^{11}C assimilate to the non-fertilized ovaries in pea within 2 h after hormonal treatment (Jahnke et al., 1989). The duration of cell division after anthesis may determine the final cell number of the fruit, particularly those with a long period of cell division after anthesis. Even in those with a moderately long cell division period, fruit size can still be altered by factors affecting the duration of cell division. For instance, in apples, cell division in the flesh (pith and cortex) ceases about 3–4 weeks after anthesis but can be prolonged by fruit thinning, resulting in bigger fruit (Sharples, 1964). The increase of cell number in the ovary in response to blossom thinning accounts for the proportional increase of the final fruit mass (Quinlan & Preston, 1968). As the increase of individual fruit mass by blossom thinning was greater than that by fruit thinning, the alteration of cell division activity before anthesis is more effective than after anthesis in determining final cell number, and hence final fruit size. Apart from increasing the assimilate supply or removing the autoinhibition from competing fruit, the cell number in GA- or IAA-induced parthenocarpic tomato fruit can also be increased by applying 4-CPA during the cell division period, in the first two weeks after anthesis, resulting in bigger fruit (Bunger-Kibler & Bangerth, 1982). It has been observed that the cell number of the proximal fruit, after delayed induction, was reduced (Bohner & Bangerth, 1988a). Most likely any delay in re-starting cell division will have reduced the final cell number, thus resulting in smaller fruit.

For cereal grains, grain mass is mainly related to the starch and protein storage in the endosperm. The dry matter accumulation in the wheat and barley grains is positively related to the number of endosperm cells (Cochrane & Duffus, 1983) as well as the number of starch granules per cell (Gleadow, Dalling & Halloran, 1982). They are either genetically determined among genotypes (Gleadow *et al.*, 1982) or morphologically determined between grains at different positions of the same spikelet (Singh & Jenner, 1982). However, both cell number and starch granule number can be reduced by water stress, particularly at high temperature during the cell-division period, 20 d after anthesis (Nicolas, Gleadow & Dalling, 1984).

Critical sink activity among sink types

Once the cell number is finalized in the fruit, the enlargement of the cells is probably determined by the mechanism controlling the cell wall extensibility and the accumulation of water and solutes, particularly sugars in the vacuole; the latter is affected by a number of metabolic activities inside the fruit. In many fruit, such as tomato and apple, which have a single sigmoid growth pattern, more than 80% of the fruit-growth period is due to cell enlargement. On the other hand, a number of drupes (e.g. peach), berries (e.g. grape) and compound fruits (e.g. raspberry) with double sigmoid growth patterns have their cell enlargement mainly in the third phase of fruit growth. Even in stone fruits such as peach, although only about 40% of the growth period is in the third phase, 80% of the fresh mass or 60% of the dry mass is accumulated during cell enlargement in this phase (Bollard, 1970). The cyclic growth, involving two active growth phases separated by a slow or zero growth phase between, is mainly due to less water accumulation and a predominant dry matter accumulation in the endocarp or seeds rather than in the flesh. Consequently, no marked cyclic pattern of dry matter accumulation is observed in these kinds of fruit. The pattern of fruit growth in terms of fresh mass or volume gain is, therefore, affected by the differential growth rates of various tissues as well as by the fate of the imported assimilate used for growth or for storage. However, the duration of fruit growth is mainly determined by the duration of slow growth in the second phase, which is sensitive to the growing conditions (Coombe, 1976). As fruit growth is mainly sustained by phloem transport of both water and nutrients, sink activity may be defined by the regulations of both the phloem transport toward the fruit and the unloading and compartmentation of imported assimilates inside the fruit.

Regulation of phloem transport for fruit growth

Swelling of a fleshy fruit is mainly due to cell enlargement, which is a function of water accumulation determined by the water relations of the plant. However, the rate of dry matter import by a fruit is determined by the sink strength, which is not sensitive to water relations (Ho, Grange & Picken, 1987). Phloem transport is important for the accumulation of assimilates as well as minerals during fruit growth, as most of the minerals, except for phloem-immobile elements such as calcium, are recycled from xylem to phloem in the leaves and then transported to the developing fruit with the assimilates (Oland, 1963). As it has been estimated that more than 85% of the water is imported by a developing tomato fruit via the phloem (Ho et al., 1987), it is not surprising that the ratio between phloem-mobile potassium and xylem-mobile calcium increases during apple fruit development (Askew et al., 1959). Under water stress, the volume and concentration of the phloem sap can change simultaneously to sustain the rate of dry matter accumulation in a tomato fruit (Ho et al., 1987). The osmotic stress of the cells was decreased by a change of the compartmentation of carbohydrate in favour of starch (Ehret & Ho, 1986a). The diurnal growth rate of fruits has been found to vary with species. Some have a higher growth rate during the day, as in tomato (Ehret & Ho, 1986b), while others are higher at night, such as apple and many fruit species (Bollard, 1970). The mid-day shrinkage of apple fruit may indicate that water can flow back from the fruit to the rest of the plant.

Unloading of assimilate

The efflux of sucrose from the seed coat to the cavity around the embryo and cotyledons has been frequently used as an example of apoplastic unloading (see Thorne, 1985). For cereals and legumes, seeds are the predominant component of the fruit and this step may be important in controlling assimilate import. The apoplastic unloading of sucrose from the seed coat into the cavity has been demonstrated to be either stimulated by high osmolarity of the bathing medium, as in soya bean and pea (Wolswinkel & Ammerlaan, 1984; Ellis & Spanswick, 1987; Grusak & Minchin, 1988), or reduced in broad bean and corn (Patrick et al., 1986; Porter, Knievel & Shannon, 1987). Similarly, inhibitors of membrane transport carriers, such as PCMBS and DNP, or proton ionophores such as CCCP, inhibit only the sucrose efflux from the seed coat in long-term experiments (Thorne & Rainbird, 1983; Wolswinkel & Ammerlaan, 1983) but not in short-term ones (Patrick et al., 1986; Grusak & Minchin, 1988). Up to now, it is uncertain whether the unloading of assimilate from the seed coat consists of both carrier-mediated and turgor-sensitive components. The unloading process inside the ovule is unlikely to be the limiting step for seed growth. Furthermore, the hydrolysis of sucrose in the cavity before the uptake by the endosperm in the corn kernel may not be essential (Cobb & Hannah, 1986). Similarly, the uptake of sugars by tomato fruit discs appears to be non-selective and not ATP dependent, and therefore may not be a controlling step (Johnson, Hall & Ho, 1988). No evidence has yet been found that the apoplastic unloading processes from the sieve tube or the uptake processes of the storage cells across the plasma membrane are the controlling steps for assimilate import into a fruit.

Conversion and accumulation of reserves

Inside the storage cells of a number of fruits, imported sugars may be stored as sugars or as starch. The regulatory processes for chemical conversion (e.g. hydrolysis of sucrose or synthesis of polysaccharides) or the physical compartmentation (e.g. storage in the plastids or in the vacuoles) may therefore control the import of assimilates into the sugar- or starch-sinks (Ho, 1986).

The synthesis of starch is the critical process determining the dry mass of the starch-sinks, such as cereal and legume grains. Sucrose is the preferable substrate; its supply does not normally limit starch synthesis in wheat grains (Rijven & Gifford, 1983; Jenner & Rathjen, 1975). It is possible that uptake of substrate or the enzyme activities in the amyloplast for starch biosynthesis are limiting. However, the substrate across the amyloplast membrane for starch synthesis may not be triose-phosphate (Jenner, 1976), but glucose-1-P and glucose-6-P, as no fructose-1,6-biphosphatase was found in the amyloplasts (Keeling *et al.*, 1988). In maize, the enzymic regulation of the biosynthesis of starch is genetically determined and regulates the import of assimilate (Koch *et al.*, 1982). The lack of critical enzymes such as ADP-glucose pyrophosphorylase in the endosperm of the *shrunken*-2 mutant may be responsible for the smaller size of kernel (Tsai & Nelson, 1966). It has been proposed that triose phosphate may be the substrate across the amyloplast membrane for starch synthesis in corn (Doehlert, Kuo & Felker, 1988).

In a great number of fleshy fruits, hexoses are the predominant stored sugars, with the exception of melon, which stores sucrose. The storage of hexoses in grape intensifies at the third phase when berry softening begins and the concentration of glucose and fructose in the fruit juice increases from less than 20 g l^{-1} to about 100 g l^{-1}. However, although apoplastic unloading may be the main route, the exact roles of invertase and ABA in the sugar accumulation of the grape berry is still uncertain (Coombe, 1989). The accumulation of hexoses by the vacuoles may be facilitated by a group translocator at the tonoplast (Brown & Coombe, 1982). In the light of the recent dispute over the existence of group translocators in sugarcane (Preisser & Komor, 1988; Maretzki & Thom, 1988), the control of sugar accumulation in the vacuole requires further investigation.

The import of sucrose by tomato fruit may be regulated by the hydrolysis of sucrose inside the fruit (Walker, Ho & Baker, 1978). However, the total activity of acid invertase of the fruit is far in excess of the hydrolysis of all the imported sucrose (Johnson *et al.*, 1988). Although imported sucrose may be partly hydrolysed in the apoplast, the main route for sucrose unloading in tomato fruit is likely to be symplastic (Damon *et al.*, 1988). On the other hand, there is sucrose synthase in the cytosol and its activity changes in parallel with the rate of assimilate import as well as the synthesis of starch, which is regulated by ADP glucose pyrophosphorylase (Robinson, Hewitt & Bennet, 1988; A. Demnitz, unpublished). Based on present knowledge, sucrose unloaded symplastically will be hydrolysed by sucrose synthase in the cytosol and the compartmentation between hexoses and starch will be regulated by the activity of ADP glucose pyrophosphorylase. In the case of apoplastic unloading, sucrose will be hydrolysed by the wall-bound acid invertase and will be taken up by the storage cell as hexoses (Fig. 6). Until the subcellular site of sucrose hydrolysis is identified, it will not be known whether the accumulation of hexoses in the vacuoles of tomato fruit is

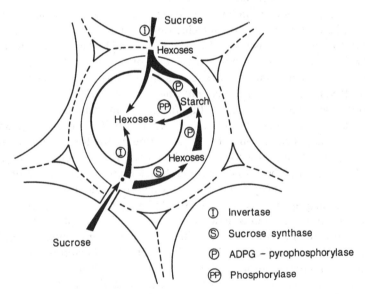

Fig. 6. The possible pathways and the enzymic regulations for the metabolism of imported sucrose in tomato fruit.

limited by the hydrolysis of sucrose in the cell wall, or in the vacuole by acid invertase, or in the cytosol by sucrose synthase, or by the uptake of sugars by the tonoplast.

Conclusions

Fruit growth is subjected to sink competition for assimilate throughout reproductive development. The priority of the reproductive organs for assimilate changes and is determined by their sink strength relative to other sinks but is independent from the availability of assimilate supply. Although the potential sink strength of a fruit is determined by sink size, such as cell number and the storage capacity within, the actual sink strength is determined by sink activity, such as processes regulating the import of water and sugars. Therefore, fruit growth is limited by both physical and physiological constraints.

For fleshy fruits, cell enlargement is the main attribute of final fruit size and is sensitive to temperature and the plant's water relations. The regulation of sugar accumulation in the vacuole is critical to cell enlargement. For grains, cell enlargement is limited and dry matter accumulation is regulated by starch synthesis in the plastids. The key to manipulating fruit growth is to understand the environmental effects on both the hormonal regulation of cell division and the genetic determination of enzymic regulation in chemical conversion (e.g. sucrose hydrolysis, starch biosynthesis) and physical compartmentation (e.g. sugar transport across tonoplast) of imported sugars during cell enlargement.

References

Abbott, D.L. (1984). *The apple tree: physiology and management.* London: Grower Books.

Askew, H.O., Chittenden, E.T., Monk, R.J. & Watson, J. (1959). Chemical investigations on bitter pit of apples. I. Physical and chemical changes in leaves and fruits of Cox's orange variety during the season. *New Zealand Journal of Agricultural Research* **2**, 1167–86.

Austin, R.B., Edrich, J.A., Ford, M.A. & Blackwell, R.D. (1977). The fate of the dry matter, carbohydrates and ^{14}C loss from the leaves and stems of wheat during grain filling. *Annals of Botany* **41**, 1309–21.

Bangerth, F. (1981). Some effects of endogenous and exogenous hormones and growth regulators on growth and development of tomato fruits. In *Aspects and Prospects of Plant Growth Regulators* (ed. B. Jeffcoat), pp. 141–50. Wantage: British Plant Growth Regulator Group.

Bangerth, F. (1989). Dominance among fruit/sinks and the search for a correlative signal. *Physiologia Plantarum* **76**, 608–14.

Bangerth, F. & Ho, L.C. (1984). Fruit position and fruit set sequence in a truss as factors determining final size of tomato fruits. *Annals of Botany* **53**, 315–19.

Barlow, H.W.B. & Smith, J.G. (1971). Effect of cropping on growth of the apple tree. *Report of East Malling Research Station for 1970*, p. 52.

Beadle, N.C.W. (1937). Studies on the growth and respiration of tomato fruit and their relationship to carbohydrate content. *Australian Journal of Experimental Biology and Medical Science* **15**, 173–89.

Bewley, W.F. (1929). The influence of bright sunshine upon the tomato under glass. *Annals of Applied Biology* **16**, 281–7.

Bohner, J. & Bangerth, F. (1988a). Effects of fruit set sequence and defoliation on cell number, cell size and hormone levels of tomato fruits (*Lycopersicon esculentum* Mill.) within a truss. *Plant Growth Regulation* **7**, 141–55.

Bohner, J. & Bangerth, F. (1988b). Cell number, cell size and hormone level in semi-isogenic mutants of *Lycopersicon pimpinellifolium* differing in fruit size. *Physiologia Plantarum* **72**, 316–20.

Bollard, E.G. (1970). The physiology and nutrition of developing fruits. In *The Biochemistry of Fruits and their Products*, vol. 1 (ed. A.C. Humes), pp. 387–425. London: Academic Press.

Bradley, M.V. & Crane, J.C. (1955). The effect of 2,4,5-Trichlorophenoxyacetic acid on cell and nuclear size and endopolyploidy in parenchyma of apricot fruits. *American Journal of Botany* **42**, 273–81.

Brooking, I.R. & Kirby, E.J.M. (1981). Interrelationships between stem and ear development in winter wheat: the effects of a Norin 10 dwarfing gene, Gai/Rht2. *Journal of Agricultural Science, Cambridge* **97**, 373–81.

Brown, S.C. & Coombe, B.G. (1982). Sugar transport by an enzyme complex at the tonoplast of grape pericarp cells. *Naturwissenschaften* **69**, 43–5.

Bunger-Kibler, S. & Bangerth, F. (1982). Relationship between cell number, cell size and fruit size of seeded fruits of tomato (*Lycopersicum esculentum* Mill) and those induced parthenocarpically by application of plant growth regulators. *Plant Growth Regulation* **1**, 143–54.

Buttrose, M.S. & Sedgley, M. (1978). Some effects of light intensity, daylength and temperature on growth of fruiting and non-fruiting watermelon. *Annals of Botany* **42**, 599–608.

Chalmers, D.J. & van den Ende, B. (1975). Productivity of peach trees: factors affecting dry weight distribution during tree growth. *Annals of Botany* **39**, 423–32.

Cobb, B.G. & Hannah, L.C. (1986). Sugar utilization by developing wild type and shrunken-2 maize kernels. *Plant Physiology* **80**, 609–11.

Cochrane, M.P. & Duffus, C.M. (1983). Endosperm cell number in cultivars of barley differing in grain weight. *Annals of Applied Biology* **102**, 177–81.

Cockshull, K.E. (1988). The integration of plant physiology with physical changes in the glasshouse climate. *Acta Horticulturae* **229**, 113–23.

Coombe, B.G. (1965). The effect of growth substance and leaf number or fruit set and size of corinth and Sultanina grapes. *Journal of Horticultural Science* **40**, 307–16.

Coombe, B.G. (1976). The development of fleshy fruit. *Annual Review of Plant Physiology* **27**, 207–28.

Coombe, B.G. (1989). The grape berry as a sink. *Acta Horticulturae* **239**, 149–58.

Cooper, A.J. (1964). A study of the development of the first inflorescence of glasshouse tomatoes. *Journal of Horticultural Science* **39**, 92–7.

Crane, J.C. (1965). The chemical induction of parthenocarpy in the calimyrna fig and its physiological significance. *Plant Physiology* **40**, 606–10.

Damon, S., Hewitt, J., Nieder, M. & Bennett, A.B. (1988). Sink metabolism in tomato fruit. II. Phloem unloading and sugar uptake. *Plant Physiology* **87**, 731–6.

Davies, J.W. & Cocking, E.C. (1965). Change in carbohydrates, proteins and nucleic acids during cellular development in tomato locule tissue. *Planta* **67**, 242–53.

de Stigter, H.C.M. (1970). Growth relations between individual fruits and between fruit and roots in cucumber. *Netherlands Journal of Agricultural Science* **17**, 209–14.

Doehlert, D.C., Kuo, T.M. & Felker, F. (1988). Enzymes of sucrose and hexose metabolism in developing kernels of two inbreds of maize. *Plant Physiology* **86**, 1013–19.

Ehret, D.L. & Ho, L.C. (1986a). Effects of salinity on dry matter partitioning and fruit growth in tomatoes grown in nutrient film culture. *Journal of Horticultural Science* **61**, 361–7.

Ehret, D.L. & Ho, L.C. (1986b). Effects of osmotic potential in nutrient solution on diurnal growth of tomato fruit. *Journal of Experimental Botany* **37**, 1294–302.

Ellis, E.C. & Spanswick, R.M. (1987). Sugar efflux from attached seed coats of *Glycine max* (L) Merr. *Journal of Experimental Botany* **38**, 1470–83.

Farineau, J. & Laval-Martin, D. (1977). Light versus dark carbon metabolism in cherry tomato fruits. II. Relationship between malate metabolism and photosynthetic activity. *Plant Physiology* **60**, 877–80.

Fisher, R.A. & Stockman, Y.M. (1986). Increased kernel number in Norin 10-derived dwarf wheat: evaluation of the cause. *Australian Journal of Plant Physiology* **13**, 767–84.

Gallagher, J.A., Biscoe, P.V. & Scott, R.K. (1975). Barley and its environment V. Stability of grain weight. *Journal of Applied Ecology* **12**, 319–36.

Gifford, R.M. & Evans, L.T. (1981). Photosynthesis, carbon partitioning and yield. *Annual Review of Plant Physiology* **32**, 485–509.

Gifford, R.M., Throne, J.H., Hitz, W.D. & Giaquinta, R.T. (1984). Crop productivity and photoassimilate partitioning. *Science* **225**, 801–8.

Gleadow, R.M., Dalling, M.J. & Halloran, G.M. (1982). Variation in endosperm characteristics and nitrogen content in six wheat lines. *Australian Journal of Plant Physiology* **9**, 539–51.

Goldwin, G.K. (1984). Factors affecting hormone-assisted setting of Cox's apple. *Acta Horticulturae* **149**, 161–71.

Grusak, M.A. & Minchin, P.E.H. (1988). Seed coat unloading in *Pisum sativum* – Osmotic effects in attached versus excised empty ovules. *Journal of Experimental Botany* **39**, 543–59.

Hansen, P. (1969). ^{14}C studies on apple trees: IV. Photosynthate consumption in fruits in relation to the leaf-fruit ratio and to the leaf-fruit positions. *Physiologia Plantarum* **22**, 186–98.

Hansen, P. (1980). Crop load and nutrient translocation. In *Mineral Nutrition of Fruit Trees* (ed. D. Atkinson, J.E. Jackson, R.O. Sharples & W.M. Waller), pp. 201–12. London: Butterworths.

Hansen, P. (1989). Source/sink effects in fruits; an evaluation of various elements. In *Manipulation of Fruiting* (ed. C.J. Wright), pp. 29–38. London: Butterworths.

Havis, A.L. (1943). A developmental analysis of the strawberry fruit. *American Journal of Botany* **30**, 311–14.

Heim, G., Landsberg, J.J., Watson, R.L. & Brain, P. (1979). The ecophysiology of apple trees: dry matter production and partitioning by young Golden Delicious trees in France and England. *Journal of Applied Ecology* **16**, 179–94.

Ho, L.C. (1979). Regulation of assimilate translocation between leaves and fruits in the tomato. *Annals of Botany* **43**, 437–48.

Ho, L.C. (1980). Control of import into tomato fruits. *Berichte der Deutschen Botanischen Gesellschaft* **93**, 315–25.

Ho, L.C. (1986). Metabolism and compartmentation of translocates in sink organs. In *Phloem Transport* (ed. J. Cronshaw, W.J. Lucas & R.T. Giaquinta), pp. 317–24. New York: Alan R. Liss.

Ho, L.C. (1988). Metabolism and compartmentation of imported sugars in sink organs in relation to sink strength. *Annual Review of Plant Physiology and Plant Molecular Biology* **39**, 355–78.

Ho, L.C., Grange, R.I. & Picken, A.J. (1987). An analysis of the accumulation of water and dry matter in tomato fruit. *Plant, Cell and Environment* **10**, 157–62.

Ho, L.C., Grange, R.I. & Shaw, A.F. (1989). Source/sink regulation. In

Transport of Photoassimilate (ed. D.A. Baker & J. Milburn), pp. 306–43. Harlow, Essex: Longman.

Ho, L.C. & Hewitt, J.D. (1986). Fruit development. In *The Tomato Crop* (ed. J.G. Atherton & J. Rudich), pp. 201–39. London: Chapman and Hall.

Ho, L.C. & Rees, A.R. (1977). The distribution of current photosynthesis to growth and development in the tulip during flowering. *New Phytologist* **78**, 65–70.

Hurd, R.G., Gay, A.P. & Mountifield, A.C. (1979). The effect of partial flower removal on the relation between root, shoot and fruit growth in the indeterminate tomato. *Annals of Applied Biology* **93**, 77–89.

Imanishi, S. & Hiura, I. (1975). Relationship between fruit weight and seed content in the tomato. *Journal of the Japanese Society of Horticultural Science* **44**, 33–40.

Jackson, D.I. & Coombe, B.G. (1966). The growth of apricot fruit. I. Morphological changes during development and the effects of various tree factors. *Australian Journal of Agricultural Research* **17**, 465–77.

Jahnke, S., Bier, D., Estruch, J.J. & Beltram, J.P. (1989). Distribution of photoassimilates in the pea plant: chronology of events in non-fertilised ovaries and effects of gibberellic acid. *Planta* **180**, 53–60.

Jenner, C.F. (1976). Wheat grains and spinach leaves as accumulators of starch. In *Transport and Transfer Processes in Plants* (ed. J.B. Passioura), pp. 73–83. London: Academic Press.

Jenner, C.F. & Rathjen, A.J. (1975). Factors regulating the accumulation of starch in ripening wheat grains. *Australian Journal of Plant Physiology* **2**, 311–22.

Johnson, C., Hall, J.L. & Ho, L.C. (1988). Pathways of uptake and accumulation of sugars in tomato fruit. *Annals of Botany* **61**, 593–603.

Keeling, P.L., Wood, J.R., Tyson, R.H. & Bridges, I.G. (1988). Starch biosynthesis in developing wheat grain. Evidence against the direct involvement of triose phosphates in the metabolic pathway. *Plant Physiology* **87**, 311–19.

Khan, A.A. & Sagar, G.R. (1966). Distribution of ^{14}C-labelled products of photosynthesis during the commercial life of the tomato crop. *Annals of Botany* **30**, 727–43.

Kinet, J.M. (1977). Effect of light condition on the development of inflorescence in tomato. *Scientia Horticulturae* **6**, 15–26.

Kinet, J.M. (1987). Inflorescence development in tomato: control by light, growth regulators, and apical dominance. *Plant Physiology (Life Science Advance)* **6**, 121–7.

Kinet, J.M., Hurdebise, D., Parmentier, A. & Stainier, R. (1978). Promotion of inflorescence development by growth substance treatment to tomato plants growth in insufficient light conditions. *Journal of the American Society for Horticultural Science* **103**, 724–9.

Kinet, J.M., Zime, V., Linotte, A., Jacqmond, A. & Bernier, G. (1986). Resumption of cellular activity induced by cytokinin and gibberellin in treatments in tomato flowers targeted for abortion in unfavourable light conditions. *Physiologia Plantarum* **64**, 67–73.

Knavel, D.E. (1969). Influence of growth retardants on growth, nutrient content, and yield of tomato plants grown at various fertility levels. *Journal of the American Society for Horticultural Science* **94**, 32–5.

Koch, K.E., Tsui, C.L., Schrader, L.E. & Nelson, O.E. (1982). Source-sink relations in maize mutants with starch-deficient endosperms. *Plant Physiology* **70**, 322–5.

Lai, R., Wolley, D.J. & Lewes, G.S. (1990). The effect of inter-fruit competition, type of fruit lateral and time of anthesis on the fruit of kiwifruit (*Actinidia deliciosa*). *Journal of Horticultural Science* **65**, 87–96.

Laval-Martin, D., Farineau, J. & Diamond, J. (1977). Light versus dark carbon metabolism in cherry tomato fruits. I. Occurrence of photosynthesis. Study of the intermediates. *Plant Physiology* **60**, 872–6.

Leonard, M., Kinet, J.M., Bodson, M. & Bernier, G. (1983). Enhanced inflorescence development in tomato by growth substance treatments in relation to ^{14}C-assimilate distribution. *Physiologia Plantarum* **57**, 85–9.

Leopold, A.C. & Lam, S.L. (1960). A leaf factor influencing tomato earliness. *Proceedings of the American Society for Horticultural Science* **76**, 543–7.

Maretzki, A. & Thom, M. (1988). High performance liquid chromatography-based re-evaluation of disaccharides produced upon incubation of sugarcane vacuoles with UDP-glucose. *Plant Physiology* **88**, 266–9.

Monselise, S.P. & Goldsmidt, E.E. (1982). Alternate bearing in fruit trees. *Horticultural Reviews* **4**, 128–73.

Nicolas, M.E., Gleadow, R.M. & Dalling, M.J. (1984). Effects of drought and high temperature on grain weight in wheat. *Australian Journal of Plant Physiology* **11**, 553–66.

Nilson, E.B., Johnson, V.A. & Gardner, C.O. (1957). Parenchyma and epidermal cell length in relation to plant height and culm internode length in winter wheat. *Botanical Gazette* **119**, 38–43.

Nooden, L.D. & Guiamet, J.J. (1989). Regulation of assimilation and senescence by the fruit in monocarpic plants. *Physiologia Plantarum* **77**, 267–74.

Oland, K. (1963). Changes in the content of dry matter and major nutrient elements of apple foliage during senescence and abscission. *Physiologia Plantarum* **16**, 682–94.

Patrick, J.W. Jacobs, E., Offler, C.E. & Cram, W.J. (1986). Photosynthate unloading from seed coats of *Phaseolus vulgaris* L. Nature and cellular location of turgor-sensitive unloading. *Journal of Experimental Botany* **37**, 1006–19.

Pearson, J.A. & Robertson, R.N. (1953). The physiology of growth in apple fruits. IV. Seasonal variation in cell size, nitrogen metabolism and respiration in developing granny smith apple. *Australian Journal of Biological Science* **6**, 1–20.

Porter, G.A., Knievel, D.P. & Shannon, J.C. (1987). Assimilate unloading from maize (*Zea mays* L.) pedicel tissues. I. Evidence for regulation of unloading by cell turgor. *Plant Physiology* **83**, 131–6.

Preisser, J. & Komor, E. (1988). Analysis of the reaction products from incubation of sugarcane vacuoles with uridine-diphosphate-glucose: No evidence for the group translocator. *Plant Physiology* **88**, 259–65.

Quinlan, J.D. & Preston, A.P. (1968). Effects of thinning blossom and fruitlets on growth and cropping of Sunset apple. *Journal of Horticultural Science* **43**, 373–81.

Rawson, H.M. & Evans, L.T. (1971). The contribution of stem reserves to grain development in a range of wheat culture of different height. *Australian Journal of Agricultural Research* **22**, 851–63.

Rijven, A.H.G.C. & Gifford, R.M. (1983). Accumulation and conversion of sugars by developing wheat grains: 3. Non-diffusional uptake of sucrose, the substrate preferred by endosperm slices. *Plant, Cell and Environment* **6**, 417–25.

Robinson, N.L., Hewitt, J.D. & Bennet, A.B. (1988). Sink metabolism in tomato fruit. I. Developmental changes in carbohydrate metabolizing enzymes. *Plant Physiology* **87**, 727–30.

Roper, T.R., Keller, J.D., Loescher, W.H. & Rom, C.R. (1988). Photosynthesis and carbohydrate partitioning in sweet cherry: Fruiting effects. *Physiologia Plantarum* **72**, 42–7.

Russell, C.R. & Morris, D.A. (1983). Patterns of assimilate distribution and source–sink relationships in the young reproductive tomato plant (*Lycopersicon esculentum* Mill.). *Annals of Botany* **52**, 357–63.

Rylski, I. (1979). Fruit set and development of seeded and seedless tomato fruits under diverse regimes of temperature and pollination. *Journal of the American Society for Horticultural Science* **104**, 835–8.

Sawhney, V.K. & Dabbs, D.H. (1978). Gibberellic acid induced multilocular fruits in tomato and the role of locule number and seed number in fruit size. *Canadian Journal of Botany* **61**, 1258–65.

Sharples, R.O. (1964). The effects of fruit thinning on the development of Cox's Orange Pippin apples in relation to the incidence of storage disorders. *Journal of Horticultural Science* **39**, 224–35.

Singh, B.K. & Jenner, C.F. (1982). Association between concentrations of organic nutrients in the grain, endsosperm cell number and grain dry weight within the ear of wheat. *Australian Journal of Plant Physiology* **9**, 83–95.

Slack, G. & Calvert, A. (1977). The effect of truss removal on the yield of early sown tomatoes. *Journal of Horticultural Science* **52**, 309–15.

Smith, P.F. (1976). Collapse of 'Murcott' tangerine trees. *Journal of the American Society for Horticultural Science* **101**, 23–5.

Stephenson, A.G., Devlin, B. & Horton, J.B. (1988). The effects of seed number and prior fruit dominance on the patterns of fruit production in *Cucurbita pepo* (zucchini squash). *Annals of Botany* **62**, 653–61.

Tamas, I.A., Wallace, D.H., Ludford, P.M. & Ozbun, J.L. (1979). Effect of older fruits on abortion and abscisic acid concentration of younger fruits in *Phaseolus vulgaris* L. *Plant Physiology* **64**, 620–2.

Tanaka, A., Fujita, K. & Kikuchi, K. (1974). Nutrio-physiological studies on the tomato plant III. Photosynthetic rate of individual leaves in relation to the dry matter production of plants. *Soil Science and Plant Nutrition* **20**, 173–83.

Teitel, D.C., Arad, S., Birnbaum, E. & Mizrahi, Y. (1985). Growth and development of tomato fruit *in vivo* and *in vitro*. *Plant Growth Regulation* **3**, 179–89.

Thomson, J.D. (1989). Development of ovules and pollen among flowers within inflorescences. *Evolutionary Trends in Plants* **3**, 65–8.

Thorne, J.H. (1985). Phloem unloading of C and N assimilates in developing seeds. *Annual Review of Plant Physiology* **36**, 317–43.

Thorne, J.H. & Rainbird, R.M. (1983). An *in vivo* technique for the study of phloem unloading in seed coats of developing soybean seeds. *Plant Physiology* **72**, 268–71.

Tsai, C.Y. & Nelson, O.E. (1966). Starch deficient maize mutant lacking diphosphate glucose pyrophosporylase activity. *Science* **151**, 341–3.

Varga, A. & Bruinsma, J. (1976). Roles of seeds and auxins in tomato fruit growth. *Zeitschrift für Pflanzenphysiologie* **80**, 95–104.

Waister, P.D. & Wright, C.J. (1986). Dry matter partitioning in cane fruit. In *Manipulation of Fruiting* (ed. C.J. Wright), pp. 51–62. London: Butterworths.

Walker, A.J., Ho, L.C. & Baker, D.A. (1978). Carbon translocation in the tomato: pathway to carbon metabolism in the fruit. *Annals of Botany* **42**, 901–9.

Watson, D.J. (1968). A prospect of crop physiology. *Annals of Applied Biology* **62**, 1–9.

Wolswinkel, P. & Ammerlaan, A. (1983). Phloem unloading in developing seeds of *Vicia faba* L. The effect of several inhibitors on the release of sucrose and amino acids by the seed coat. *Planta* **158**, 205–15.

Wolswinkel, P. & Ammerlaan, A. (1984). Turgor-sensitive sucrose and amino acid transport into developing seeds of *Pisum sativum*. Effect of a high sucrose or mannitol concentration in experiments with empty ovules. *Physiologia Plantarum* **61**, 172–82.

C. M. DUFFUS

Control of grain growth and development

Introduction

Seeds contribute on a world basis some 55% of our daily per capita protein and energy supply. Of this, 90% is accounted for by cereals (Duffus & Slaughter, 1980). The contribution is even greater if allowance is made for the inclusion of cereals and other seed crops in the diet of farm animals. It is therefore not surprising that there has long been an interest in the mechanisms regulating seed growth and maturation, since such studies could lead to the identification of key factors involved in the control of yield and quality in the harvested material.

The aim of this chapter is, then, to review present understanding of the control mechanisms involved in seed growth and development. The major emphasis is on the cereals, including barley, wheat, rice, oats and maize. The morphological, physiological and biochemical changes accompanying grain growth and development are described, and the influence of genotype and environmental conditions on these changes is discussed.

Developmental morphology

The cereal seed is surrounded by a seed coat or testa, which is fused to the pericarp, forming a single-seeded fruit known as a caryopsis. The harvested material is termed a grain or kernel; in wheat, rye, triticale, maize, grain sorghum and the naked-grain millets, this consists solely of a caryopsis. In oats, barley, rice and the husked types of sorghum and millet, the grain consists of a caryopsis, together with the lemma and palea, which adhere to the outer surface. Some naked barleys are found, for example cvs Himalaya and Nakta. These find a particular use in germination studies, where the absence of a husk allows convenient isolation of aleurone and other cells. In wheat, oats, rye, barley, triticale and rice, there is a furrow or crease, deep in wheat, shallow in barley, which runs along the ventral side of the grain. The embryo is present on the opposite, or dorsal, side at the base of the caryopsis, and is effectively surrounded by endosperm. The endosperm itself accounts for the major part of the starch and protein reserves of the harvest-ripe grain.

The literature relating to the developmental morphology of cereal caryopses has been reviewed previously (Duffus & Cochrane, 1982; Duffus, 1987). In this account therefore, some emphasis will be put on more recent progress in our understanding of the morphological changes accompanying caryopsis development.

The endosperm

The events following fertilization have been well described for endosperm by Brink & Cooper (1947). Of the two male nuclei from the pollen tube, one fuses with the female egg nucleus to form the diploid zygote. The other male nucleus fuses with the two polar nuclei of the embryo sac to form the triploid endosperm nucleus. The pattern of grain filling and maturation that follows depends on environment, species and cultivar. For example, the period between anthesis or pollination and harvest-ripeness can be over 60 d for barley and wheat grown in Scotland. In Australia or Canada, on the other hand, this period may be as short as 40 d. Thus, unless environmental conditions are specified or known, age in days after anthesis is not a good indicator of stage of morphological development. However, in the absence of a generally accepted set of morphological markers, days after anthesis is generally taken as a measure of grain development.

Division of the endosperm nucleus precedes that of the zygote; initially, there is a period of free nuclear division in the endosperm followed a few days later by endosperm cellularization. Early divisions take place throughout the endosperm, but later divisions are confined to the outermost layer of cells, which ultimately become the aleurone layer (Kvaale & Olsen, 1986). There are no divisions in the outer layer of endosperm cells at the crease. Aleurone formation is of particular interest: these are the cells that develop the alpha amylase responsible for starch degradation during malting. Barley seems to be particularly well adapted to this purpose since, depending on cultivar, the aleurone layer may be three to six cells thick whereas in wheat and oats it is one cell thick. Differentiation of aleurone cells can begin before cell division ends (Morrison, Kuo & O'Brien, 1975) and hence the final number of aleurone cells will depend on whether a daughter cell de-differentiates and becomes a starchy endosperm cell or remains as an aleurone cell.

Once cellularization is complete, the starchy endosperm cells expand considerably; this stage of development is associated with a rapid synthesis and deposition of starch and protein. There is some controversy about the number of days after anthesis at which cell division terminates. Part of the problem stems from the possibility that there may be a small but significant rate of cell turnover as a result of endosperm digestion in the modified cells next to the scutellum (Smart & O'Brien, 1983). If cell numbers are estimated by counting nuclei, then cell division cannot necessarily be detected by this method because, even if cell numbers reach a constant value, cell division may still be continuing. In wheat, maximum cell number was recorded by 14–20 d (Briarty, Hughes & Evers, 1979); more recent work (Huber & Grabe, 1987) has reported that the last nuclear divisions were observed between the twelfth and fourteenth days post-anthesis. In barley aleurone, in contrast, mitotic figures can be seen as late as 32 days post anthesis or rather more than half way through the period of grain development (Cochrane & Duffus, 1981; Kvaale & Olsen, 1986).

It should be emphasized that allowance must be made for genetic, environmental and experimental variation when comparing developmental processes. Total endosperm cell numbers may differ markedly between species (Palmer, 1989). Environmental variables, such as temperature, irradiance and water supply, may affect the rate of endosperm cell division (Duffus & Cochrane, 1982). Finally, isolated endosperms may be either contaminated with testa, or stripped of their aleurone cells, depending on the stage of development at which samples are taken.

Starch deposition takes place within amyloplasts (Fig. 1); these are visible in the endosperm even before cellularization is complete. The first-formed or A-type starch granules grow steadily within the amyloplasts, reaching a maximum diameter of around 45 µm. In wheat and barley, these are round or lenticular in shape with a peripheral groove. A second population of small or B-type amyloplasts appear later. These contain spherical starch granules, which eventually grow to a maximum size of 10 µm in diameter. Although relatively few in number, the greater part of endosperm starch is contained in the A-type granules (Duffus & Murdoch, 1979; Chojecki, Gale & Bayliss, 1986). In wheat, it appears that the B-type granules are initiated in the stroma of A-type amyloplasts, often in associaton with the peripheral groove and its associated tubuli (Parker, 1985). The A-type amyloplasts also contain plastid DNA (Day, Bayliss & Pryke, 1987). The starches of wheat, barley, rye and triticale are associated with lysophospholipids (Morrison, 1988); it may be that these are somehow involved in the regulation of starch synthesis and deposition.

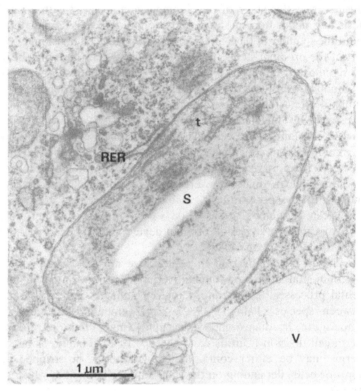

Fig. 1. Amyloplast in a sub-aleurone cell from a mid-grain transverse section of a grain of barley cv. Midas, fixed 19 d after anthesis. The barley was grown in field conditions where the time from anthesis to harvest-ripeness was 60 d. S, starch; t, tubuli; RER, rough endoplasmic reticulum; V, vacuole. (Micrograph from M. P. Cochrane.)

As development proceeds, the storage protein of the developing endosperm is deposited in membrane-bound structures called protein bodies (Fig. 2). Their form and origin seems to depend on the state of differentiation of the cell concerned. For example, if the outermost cells of the starchy endosperm are fixed for electron microscopy during the early stages of differentiation, small deposits of storage protein are observed in large vacuoles. However, in older starchy endosperm cells of barley, storage protein is found in structures that appear to have been derived from dilations of the endoplasmic reticulum. These membrane-bound deposits form aggregates surrounded by layers of membranes (M.P. Cochrane, personal communication). Bechtel (1985), in a generalized

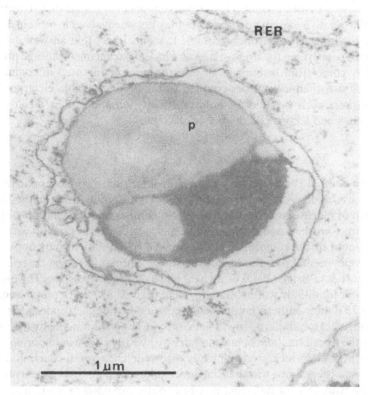

Fig. 2. Protein deposit in a starchy endosperm cell from a mid-grain transverse section of a grain of barley cv. Midas, fixed 25 d after anthesis. The barley was grown in field conditions where the time from anthesis to harvest-ripeness was 60 d. RER, rough endoplasmic reticulum; p, protein deposit. (Micrograph from M. P. Cochrane.)

scheme for wheat, has proposed that storage proteins are initially formed in the rough endoplasmic reticulum and processed via dictyosomes into vesicles, which then enlarge into protein bodies.

Aleurone-cell development and formation differs in a number of ways from that of the starchy endosperm. That is, by maturity in wheat, any starch present has disappeared, the cell walls have become thick and lipid bodies have become abundant. Protein is deposited only in the later stages of development (Morrison *et al.*, 1975) and protein-body structure is quite distinct from that of the starchy endosperm. The amount of protein per cell is also higher than in the starchy endosperm (Kent, 1983). In oat aleurone (Peterson, Saigo & Holy, 1985) phytin is deposited within vacuoles before protein. These workers also suggested that, because of their relative scarcity, dictyosomes may not be involved in protein deposition within oat aleurone vacuoles. It seems likely that answers to questions concerning the origins, movement and packaging of storage proteins may come from techniques of electron microscopy used in conjunction with cytochemical probes coupled to readily identifiable electron-dense markers such as colloidal gold or silver. Suitable probes include antibodies and lectins (Baldo & Lee, 1987).

The embryo

The embryo is derived from the zygote after such morphogenetic processes as cell division, cell elongation, cellular differentiation and the formation of meristems. The mature embryo consists of a scutellum, which lies against crushed and depleted endosperm cells, a radicle with a rootcap covered by the coleorhiza, and a plumule covered by the coleoptile. The developmental morphology of the wheat embryo has been described by Smart & O'Brien (1983) using light and electron microscopy. The embryo develops initially within a small pouch, derived from the embryo sac, and by 15 d after anthesis is completely covered by modified endosperm cells. Nutrients for embryo growth appear to be supplied initially by hydrolysis of the nucellar parenchyma and later by hydrolysis of neighbouring endosperm cells, which are completely digested. The overall conclusion from these observations was that the developing wheat embryo appears to be a powerful sink for nutrients supplied to the caryopsis as a whole.

The ultrastructural changes observed during barley embryo development have been described by Ahluwalia (1980). At the early stages, cells in the root region were thin-walled with large vacuoles and a dense cytoplasm containing numerous ribosomes and mitochondria with few cristae. By 27 d post-anthesis, lipid bodies and amyloplasts were present and protein deposits could be seen within vacuoles. Numerous mitochondria with many cristae were also present. Nutrients apparently enter the embryo by an apoplastic route, since no plasmodesmata have been observed in the outer walls of the zygote nor in the outer epidermal walls of the embryo.

The outer layers of the caryopsis

The structure of the outer layers (Fig. 3) of developing cereal grains has been described previously (Cochrane & Duffus, 1979; Duffus, 1987). The nucellus, which surrounds the original embryo sac and lies beneath the testa, degenerates to leave a nucellar epidermis except on the ventral side

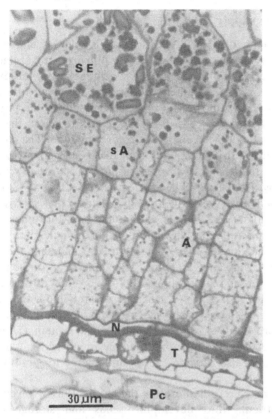

Fig. 3. Transverse section of the outer layers of a grain of barley cv. Midas, cut at mid-grain and stained in toluidine blue. The grain was glasshouse-grown; its age was estimated to be 33 d after anthesis on a developmental scale having 60 d between anthesis and harvest-ripeness. Pc, pericarp cross cell; T, testa; N, nucellus; A, aleurone; sA, sub-aleurone; SE, starchy endosperm. (Micrograph from M. P. Cochrane.)

at the crease, where it forms a mass of cells projecting into the endosperm. Surrounding the testa is the pericarp, whose inner cells contain chloroplasts. The chloroplast-containing layer is one cell thick in oats, rye and wheat, and 2–3 cells thick in barley. The outer pericarp becomes transparent in the middle stages of grain development, partly perhaps as a result of hydrolysis of the starch granules present in these cells. The cuticular layers on the outside of the nucellus and testa are continuous round the grain except in the region of the micropyle and along the crease in caryopses such as wheat or barley.

Assimilate supply

Assimilates are supplied to the caryopsis via a single vascular bundle, which terminates at the base of the pericarp in maize and sorghum. In wheat, barley, oats, rye and rice (Duffus & Cochrane, 1982) this extends along the crease; it is generally believed that the assimilates then pass through the chalazal cells in the gap between the ends of the testa and enter the endosperm via the cells of the nucellar projection, the endosperm cavity and the modified cells of the crease aleurone. The composition of pure phloem sap from wheat has been analysed using sap collected from the stylets of the small brown plant hopper (*Loadephax striatellus* Fallen) severed by a YAG laser beam (Hayashi & Chino, 1986). This work has confirmed what was only suspected before: that sucrose, at a concentration of 251 mM, is a major carbon source. More unexpected, however, is the observation that the concentration of total amino acids at 262 mM is somewhat higher than that of sucrose and that the major components are aspartate and glutamate. The amino acids of the phloem are not present in the same proportions as those in cereal protein (Table 1). For example, the percentage of aspartate is rarely more than 5% in wheat protein. Hence there will be much metabolic transformation of amino acids before their incorporation into grain protein (Duffus & Rosie, 1978). It seems likely, therefore, that any amino acids not used for protein synthesis may also provide carbon and energy for grain growth and development.

There has been much recent interest in the mechanisms involved in the transport of nutrients into developing seeds and storage organs (Wolswinkel, 1988; Ho, 1988). For example, assimilate unloading from maize pedicel tissue has been studied by using the empty seed technique (Porter, Knievel & Shannon, 1987). In this method, the distal halves of kernels are excised and the endosperm removed such that each base forms a cup attached to the ear. These cups consist of pericarp, vascular and pedicel parenchyma tissue so that assimilate unloading can be studied. The results suggest that unloading is inhibited by solute concentration in the free space. It may be, then, that the rate of assimilate removal and conversion to storage material may in turn regulate unloading. Similar experimentation is more difficult with the smaller-grained cereals. However, from measurements of assimilate concentrations in sieve tube and phloem exudates and cavity sap it has been found (Fisher & Gifford, 1986) that, in wheat, sucrose concentration in the sieve tubes was almost ten times that in the endosperm cavity sap. It was concluded that movement across the crease tissues may be an important control point in grain filling.

Table 1. *Amino acid composition (%) of total wheat-grain protein and phloem-sap amino acids*

Amino acid	Grain protein	Phloem sap
Alanine	3.3	2.6
Arginine	4.0	2.3
Aspartic acid	4.7	19.7
Asparagine	—[a]	3.9
Cystine + cysteine	2.6	—[a]
Glutamic acid	33.1	30.3
Glutamine	—[a]	3.7
Glycine	3.7	0.7
Histidine	2.2	1.9
Isoleucine	3.8	4.1
Leucine	6.7	4.7
Lysine	2.3	4.7
Methionine	1.7	0.7
Phenylalanine	4.8	3.2
Proline	11.1	—[a]
Serine	5.0	5.7
Threonine	2.8	4.9
Tryptophan	1.5	—[a]
Tryosine	2.7	2.2
Valine	4.4	4.7

[a]No data available.
Source: Data for protein from Kent (1983); data for amino acids recalculated from Hayashi & Chino (1986).

However, Jenner & Rathjen (1977), using cultured endosperms, and Lingle & Chevalier (1984), using developing barley kernels, have concluded nevertheless that the accumulation of dry matter is limited by processes within the endosperm and not by the availability of assimilates. Furthermore, the reduction in barley endosperm starch observed at elevated temperatures does not appear to be due to limiting assimilate levels but rather to decreased activity of one of the enzymes of sucrose utilization (MacLeod & Duffus, 1988a,b).

In maize (Porter *et al.*, 1985) but not in wheat (Donovan *et al.*, 1983), incoming sucrose may be hydrolysed to hexose during transport into the endosperm. However, it has been shown in maize that a sucrose analogue, 1'-fluorosucrose, can be taken up into endosperm without invertase hydrolysis (Gougler Schmalstig & Hitz, 1987) and it was concluded that hexoses are therefore not an obligatory requirement for sucrose uptake. This, together with evidence from short-term transport studies (Griffith, Jones & Brenner, 1987a), suggests that sucrose uptake into the endosperm may be a passive process. Thus in maize, as in the temperate cereals, the capacity for assimilate utilization, rather than the supply of assimilate, may regulate grain growth. Of course, when assimilate supply is low, grain growth will be limited directly (Cobb et al., 1988).

It is interesting to speculate whether or not termination of grain growth is a result of reduction in assimilate supply. Certainly Jenner & Rathjen (1977) have shown, by using wheat endosperms in culture, that incorporation of ^{14}C into compounds other than starch takes place after starch synthesis has ceased. Data from experiments in which [U-^{14}C]sucrose was supplied to detached barley ears in liquid culture shows that label moved into the grains at all stages of development up to the final dehydration stage (Cochrane, 1985). Thus it would seem that termination of grain growth is not due to a reduction in assimilate supply. It was suggested that blockage of xylem elements with pectinaceous material might be responsible. It was not possible from these results to discover whether the reduced assimilate supply observed with increasing maturity was a result of lignification and other changes observed in the cells of the chalazal region (Cochrane, 1983).

Developmental physiology and biochemistry

Biosynthesis of starch in the developing endosperm

Most of the sucrose entering the endosperm is presumably converted to starch within the developing amyloplasts. Other starch precursors will include carbon skeletons derived from amino acid metabolism and hexoses, notably in maize. The regulation of starch synthesis has recently been reviewed by Preiss (1988). In cereals (Duffus, 1987) it seems that incoming sucrose is first converted to UDP-glucose and fructose by UDP-dependent sucrose synthase:

$$\text{sucrose} + \text{UDP} \rightleftharpoons \text{UDP-glucose} + \text{fructose}.$$

Some ADP-glucose may also be produced by an ADP-dependent sucrose synthase. Since it appears that the greater part of α-(1→4)-D-glucan synthesis is catalysed by an ADP-glucose-dependent starch synthase, it has been suggested that the UDP-glucose formed, together with any hexose phosphates derived from hexoses, is first converted to glucose-1-phosphate and then to ADP-glucose via the enzyme ADP-glucose pyrophosphorylase. Starch would then be formed in a coordinated reaction involving starch synthase and branching enzyme. A de-branching enzyme might also be required to produce primers for the starch synthase reaction.

Control of starch synthesis may be exerted by the permeability of the amyloplast inner membrane as well as by the regulatory properties of the various enzymes involved. However, the form in which carbon enters the amyloplast has long been subject to speculation and hence the exact biochemical pathway, and its intracellular location, whereby sucrose is converted to starch is still unknown. One of the problems is that there are enormous difficulties in obtaining good yields of intact, undamaged, uncontaminated, starch-containing amyloplasts that can be used routinely for transport studies. Nevertheless, Entwistle & ap Rees (1988), using an amyloplast preparation from wheat endosperm, and Keeling *et al.* (1988), using ^{13}C-labelled hexoses supplied *in vivo* to intact wheat plants, have independently reached the conclusion that a hexose phosphate is the most likely candidate for entry to the amyloplast. The amyloplasts were prepared by density gradient fractionation of endosperm protoplast lysates. The absence from them of significant fructose-1,6-bisphosphatase and of pyrophosphate : fructose-6-phosphate 1-phosphotransferase ($PFK(PP_i)$) activity led to the conclusion that triose phosphate could not be converted to hexose phosphate at the rate required to sustain starch synthesis, and that therefore carbon from starch synthesis did not enter the amyloplast as triose phosphate. From the ^{13}C-labelling experiments it was found that the pattern of partial distribution of label between carbons 1 and 6 of glucose recovered from endosperm starch was similar to that found in the glucosyl and fructosyl moieties of sucrose extracted from the same tissue. It was therefore suggested that both sucrose and starch were synthesized from a common pool of intermediates such as hexose phosphates and that glucose-1-phosphate, glucose-6-phosphate or fructose-6-phosphate is the most likely candidate for entry into the amyloplast. In contrast, Echeverria *et al.* (1988), using maize kernel amyloplasts (also prepared by using density-gradient fractionation of endosperm protoplast lysate) found that triose phosphate was the preferred substrate for uptake and that amyloplasts contained all enzymes necessary to convert triose phosphates into starch. Previously it had been supposed, in the absence of any direct evidence and by analogy with chloroplasts, that carbon was transported into the amyloplast as triose phosphate (Duffus, 1984). Whether or not there are real differences between wheat and perhaps the other temperate cereals and maize, may be resolved by developing further the *in vivo* labelling techniques of Keeling *et al.* (1988). Some of the enzyme-catalysed reactions that could be involved in the conversion of sucrose to starch are shown in Fig. 4. The possible origin of the pyrophosphate required for the cytosolic pyrophosphorylase reaction has also been

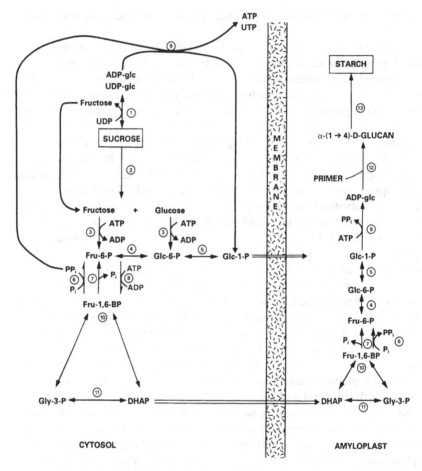

Fig. 4. Hypothetical scheme for conversion of sucrose to starch in developing amyloplasts. (1), sucrose synthase; (2), invertase; (3) hexokinase; (4), phosphohexoisomerase; (5), phosphoglucomutase; (6), PP$_i$-dependent phosphofructokinase; (7), fructose-1,6-biphosphatase; (8), ATP-dependent phosphofructokinase; (9), ADP-glucose and UDP-glucose pyrophosphorylases; (10), aldolase; (11), triosephosphateisomerase; (12), starch synthase; (13), branching enzyme. UDP-glc, UDP-glucose; ADP-glc, ADP-glucose; Glc-6-P, glucose-6-phosphate; Glc-1-P, glucose-1-phosphate; Fru-1-P, fructose-1-phosphate; Fru-6-P, fructose-6-phosphate; Fru-1,6-BP, fructose-1,6-biphosphate; Gly-3-P, glyceraldehyde-3-phosphate; DHAP, dihydroxyacetone phosphate; PP$_i$, pyrophosphate; P$_i$, inorganic phosphate.

subject to recent speculation. It may be derived from cytosolic PP_i-dependent phosphofructokinase as indicated, or alternatively from the amyloplast-located pyrophosphorylase via an adenylate–PP_i exchanger (Doehlert, Kuo & Felker, 1988).

Biosynthesis of protein and lipid in the developing endosperm

It seems likely that the proteins synthesized during the early stages of cell division and enlargement are mainly enzymes and structural proteins. The storage proteins appear later. Of these proteins, the prolamins, which are soluble in water–alcohol mixtures, appear to exert a key influence on the technological characteristics of the mature wheat grain. For example, gliadin and glutenin are involved in determining bread-making quality in wheat. The deposition of glutenins has been studied by Alldrick (1986) who showed that while glutenins were synthesized at a low constant rate from as early as 10 d post-anthesis, rates of incorporation of amino acids into high-molecular-mass glutenins reached a maximum only at 50 d post-anthesis. This correlates with the general observation that protein deposition is a relatively late event in grain development. Interestingly, however, the mRNAs encoding α-, β- and γ-type gliadins are detectable as early as 3 d post-anthesis (Reeves, Krishnan & Okita, 1986) but the gliadins themselves were detectable only by 10 d. Other workers (Tercé-Laforgue, Sallantin & Pernollet, 1987) report even earlier appearance of gliadins. Any delay in gene expression may be the result of additional control at the translational level. In developing barley endosperm, deposition of the prolamin hordein, which has adverse effects on starch dissolution during mashing, takes place several days after the corresponding mRNA species can be detected (Rahman *et al.*, 1984). It is surprising that, in spite of the enormous economic importance of protein, the control of its synthesis and deposition has received consistently less attention than that of carbohydrate metabolism and starch accumulation. This may be a function of difficulties encountered in the identification of metabolic intermediates as well as the many different storage proteins. The introduction of labelled precursors under controlled conditions could be made easier by the use of *in vitro* systems such as ear and grain culture. Maize endosperm cell cultures have been used (Lyznik & Tsai, 1989) to study protein synthesis, and although these resemble only in part the original endosperm they may prove to be a useful model system.

The greater part of grain lipid is located in the embryo and aleurone. The major class of lipid synthesized is triglyceride, although polar lipids are formed first (Weber, 1969). Cereal endosperm starch is unusual among plants in having lipids present within the granules; it is assumed (Morrison, 1988) that these lipids are complexed with amylose. It has been suggested (Vieweg & de Fekete, 1980) that this association prevents branching and breakdown, but not elongation, of amylose and that amylopectin is formed from amylose chains that are not complexed.

The embryo

Since the immature cereal embryo is capable of independent growth, even at a relatively undifferentiated stage, when cultured in a suitable medium (Cameron-Mills & Duffus, 1977), it may be assumed that the processes required for the synthesis of storage materials are present in the embryos themselves. The source of nutrients for embryo growth is unknown but is assumed to be the endosperm. Results from *in vitro* culture show that sucrose accumulation in developing embryos is driven by an active sucrose transport system (Cameron-Mills & Duffus, 1979; Griffith *et al.*, 1987*b*). This provides a mechanism whereby the embryo can compete effectively with the endosperm for sugar. The polypeptide components of the sucrose carrier in the plant plasma membrane are currently being characterized (Lemoine *et al.*, 1989).

The maturing embryo appears to differ in its sucrose relationships from the associated endosperm at the same stage of development (Duffus & Binnie, 1990). For example, the wheat embryo accumulates sucrose to concentrations around ten times higher than in the endosperm (Fig. 5). Similar observations were obtained for developing maize embryos (Griffith *et al.*, 1987*b*). This would explain the existence of an energy-dependent mechanism for sucrose uptake. The uptake of sucrose by developing maize embryos is influenced by osmotic environment; for example, it has been shown that sucrose efflux decreases with increasing mannitol concentration in the bathing medium (Wolswinkel & Amerlaan, 1989). It was therefore suggested that there might be a direct effect of apoplast solute concentration on the sucrose pumping mechanism. The levels of sucrose supplying the embryo have been implicated in the control of premature germination on the ear (Duffus, 1990). That is, germination of isolated immature embryos is suppressed by high sucrose concentrations. This may be related to the desiccant effect of sucrose where water is drawn out from the embryo, decreasing the water potential and preventing germination. *In vivo*, it may be that the high concentration of sucrose and relatively low water potential in the embryo suppresses germination. Should the sucrose pump cease to function, then the water potential of the embryo may increase and germination could then be triggered.

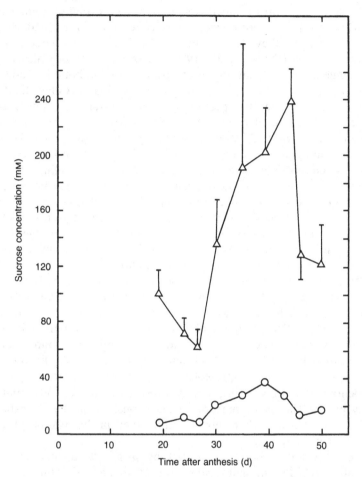

Fig. 5. Changes in sucrose concentration (mM) of developing embryos (triangles) and endosperms (circles). Each result is the mean of three determinations. Bars represent confidence limits at 95%. For all ages of endosperm, confidence limits for each determination were less than 4.4 mM. (Data from Duffus & Binnie (1990).)

The outer layers of the caryopsis

With the exception of maize, the caryopses of immature cereal grains are capable of light-dependent carbon dioxide fixation (Evans & Rawson, 1970). This is a consequence of photosynthetic activity in the chloroplast-containing cells of the pericarp (Duffus, Nutbeam & Scragg, 1985). These cells, which are often referred to as the green layer or cross cell layer, are capable of high rates of light-dependent oxygen evolution (Nutbeam & Duffus, 1978) and there is evidence to suggest that, in oats and wheat as well as barley, they have some of the features associated with C_4 photosynthesis (Duffus *et al.*, 1985). In experiments in which oxygen exchange by intact grains of wheat was measured in both light and dark over the period of grain development, it was found (Caley, Duffus & Jeffcoat, 1990) that the period in which the rate of photosynthesis exceeded the rate of respiration extended from just after anthesis to about mid-way through grain development. Outside this period there was net oxygen uptake. It appears, therefore, that the green-layer cells may make some contribution to the overall economy of grain filling. However, the source of the carbon dioxide and the fate of its photosynthetic products within the developing grain are open to speculation.

That the transparent layer of pericarp may limit oxygen uptake was suggested by the observation that respiration rates were considerably greater following its removal (Fig. 6). Since total photosynthesis is also increased, it may be that this layer additionally limits oxygen efflux (Caley *et al.*, 1990). There may also be limitations to carbon dioxide exchange. Nevertheless, it has been shown (Watson & Duffus, 1988) that immature, detached caryopses of barley and wheat are capable of fixing externally supplied $^{14}CO_2$ in both light and dark and of transferring some of the labelled material to the endosperm–embryo. Removal of the transparent layer results in a substantial increase in $^{14}CO_2$ fixation, suggesting that, despite the presence of stomata in the pericarp epidermis (Cochrane & Duffus, 1979), this layer restricts carbon dioxide uptake. It may be that the major function of the pericarp is in the re-fixation of internally derived carbon dioxide rather than in the fixation of atmospheric carbon dioxide.

This hypothesis has been tested by using ^{14}C-labelled caryopses obtained by a 15 min light incubation in $^{14}CO_2$ and subjecting them to a 3 h chase in $^{12}CO_2$ in either light or dark (Watson & Duffus, 1991). The results suggest that photosynthesis in the pericarp green layer can prevent losses of internally produced carbon dioxide, since three times as much radiocarbon remained in the caryopses incubated in the light as in the dark.

Depending on a number of assumptions, it has been estimated that, of the total endosperm starch, about 2% could be derived from fixation of atmospheric carbon dioxide (Watson & Duffus, 1988) and very much less than that (Watson & Duffus, 1991) from fixation of internally derived carbon dioxide.

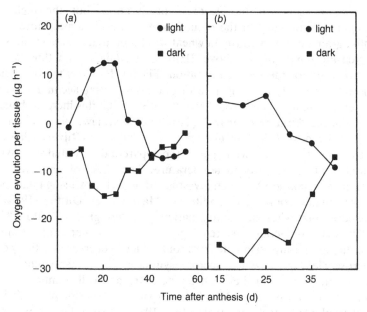

Fig. 6. Oxygen evolution for tissue in the light (circles) and in the dark (squares) during grain development in wheat cv. Sicco. (*a*) Intact caryopses. Mean standard errors are $\pm$ 1.9 μg h^{-1} (light) and $\pm$ 4.8 μg h^{-1} (dark). (*b*) Caryopses with transparent layer of pericarp removed. Mean standard errors are $\pm$ 2.4 μg h^{-1} (light) and $\pm$ 3.5 μg h^{-1} (dark). Net oxygen evolution shows that production of oxygen by photosynthesis is exceeding its use in respiration. The reduction in net photosynthesis seen when the transparent layer is removed (comparing oxygen evolution in (*a*) and (*b*)) is largely due to increased respiration, as recorded in the dark. Total photosynthesis is seen to be greater in the caryopses with the transparent layer removed (by comparing the difference between oxygen evolution in light and dark in the presence and absence of the pericarp transparent layer). (Modified from Caley *et al.* (1990).)

Environmental effects

The rate and duration of grain growth and maturation are affected by the prevailing environmental conditions. Several groups of workers (see for example Nicholas, Gleadow & Dalling, 1984) have studied the effect of elevated temperatures on grain-filling and have found that, while the rate of dry matter accumulation is increased, senescence is hastened and the duration of grain growth reduced. To a certain extent, the increase in rate can compensate for the decrease in duration, but the higher the temperature the greater the reduction in final grain mass. More specifically, grain mass at maturity in wheat was found to decrease steadily with increasing temperatures above 21/16 °C (day/night) and thus failed to compensate for the reduced duration (Fig. 7). Rice, on the other hand, showed an increase in the rate of grain dry-matter accumulation with increase in temperature from 24/19 °C to 30/25 °C; this increase balanced the reduced duration of growth (Tashiro & Wardlaw, 1989). It was suggested, from work with incubated endosperm slices (Rijven, 1986), that the basis of this difference might be the apparent differential sensitivity of starch synthase activity to temperature. The effect of temperature on starch accumulation has been investigated in barley by using intact plants grown under controlled conditions (MacLeod & Duffus, 1988a). It appeared that when developing barley ears were grown at 20/15 °C and 30/25 °C the reduction in starch deposition at the higher temperature was not due to limiting assimilate levels but rather to decreased activity of the sucrose cleavage enzyme, sucrose synthase. In corresponding experiments, no correlation between temperature and cell number could be detected over the range 15–30 °C. In similar experiments with intact barley plants (MacLeod & Duffus, 1988b) two other responses to elevated temperature were observed: (i) the volume available for starch accumulation was decreased as a result of reduction in endosperm volume; and (ii) the numbers of both A- and B-type starch granules were reduced, without a significant increase in volume, thereby decreasing the total amount of starch accumulated. Bhullar & Jenner (1986), using wheat endosperms cultured *in vitro*, confirmed that starch synthesis had a comparatively low temperature optimum and that elevated temperature led to an irreversible reduction in the capability of the endosperm to convert sucrose to starch. It is clear, however, that we have some way to go before the fundamental basis of the temperature effect is fully understood.

Interrelated with temperature is the effect of water deficit on grain-filling. However, it has been shown (Barlow *et al.*, 1980) that the grain is largely protected from water stress under drought conditions. For example, when water was withheld 10 and 20 d after anthesis, grain water

---

Fig. 7. The effect of temperature on grain development in wheat (open squares) and rice (filled squares). Grains were harvested from floret B in the central four spikelets of the ear of wheat and from the fourth, fifth and sixth spikelets from the apex of the upper two primary branches of the panicle of rice. (A) Dry mass per grain at maturity; (B) rate of grain growth; (C) duration of grain growth (from Tashiro & Wardlaw, 1989).

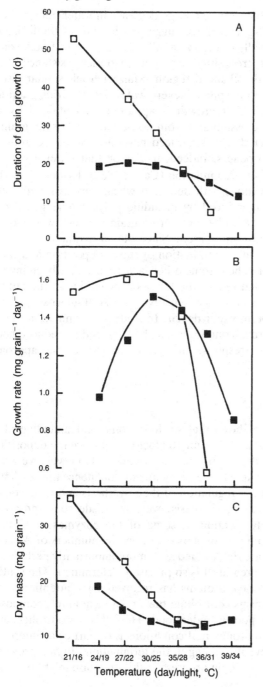

potential changed little despite a large decrease in water potential of the glumes, rachis and flag leaf. It was suggested that this hydraulic isolation might be related initially to the xylem discontinuity in the floral axis and, in longer-term water stress situations, to an increased resistance to water flow in the cells of the chalazal region. Maize kernels similarly sustain normal rates of growth in spite of severe water deficit during grain-filling (Ouattar *et al.*, 1987) but growth is apparently maintained by a re-mobilization of stored assimilates from non-grain parts to the grain.

Of course, a water deficit is induced naturally within the grain as it approaches harvest-ripeness. Indeed, it may be that water loss triggers cessation of dry matter accumulation (Caley, 1986). Evidence for this is derived from observations with detached wheat ears in liquid culture, where water loss was induced by including polyethylene glycol in the culture medium. In the endosperms from grain in which water loss was induced, starch deposition ceased. Since starch synthase activity was maintained until well after termination of starch deposition and was not affected by the water deficit, some other factor must have been involved. The results showed that one of the factors contributing to the failure to synthesize starch was the high sensitivity of ADP-glucose pyrophos-phorylase to endosperm water deficit. Indeed, the termination of starch deposition under normal conditions may be triggered by reduced activity of this enzyme in response to grain drying as it approaches harvest-ripeness.

Conclusions

The discussion above shows that we have some understanding of how physical, biological and environmental factors influence the deposition of reserve materials in developing cereal caryopses. However, we are still not in a position to be able to identify with any confidence the key limiting enzymes involved in the regulation of their synthesis and deposition.

In relation to starch biosynthesis, we have made some progress in discovering the likely location of some of the enzymes that may be involved. We still do not know however how the numbers of A- and B-type starch granules are determined and, more importantly perhaps, how the composition of starch itself is so precisely determined. The synthesis and deposition of storage proteins has received some attention but the fundamental mechanisms controlling their relative amount, composition and quality are still poorly understood. How this is controlled at the molecular level by environmental conditions is of particular interest.

The sequence of events leading to the termination of grain growth has always been a mystery. Some progress in this direction might be made if the factor initiating water loss could be identified. In turn, such work could lead to a fuller understanding of such problems as pre-harvest sprouting in wheat, and dormancy in barley.

References

Ahluwalia, B. (1980). *Mineral elements and embryo development in barley*. Ph.D. thesis, University of Edinburgh, Scotland.

Alldrick, S.P. (1986). *Protein deposition in wheat grains*. Ph.D. thesis, University of Nottingham, England.

Baldo, B.A. & Lee, J.W. (1987). Lectins as cytochemical probes of the developing wheat grain. VII. Ultrastructural location of protein bodies and nucellar epidermis with lectin-gold complexes. *Journal of Cereal Science* **6**, 211–18.

Barlow, E.W.R., Lee, J.W., Munns, R. & Smart, M.G. (1980). Water relations of the developing wheat grain. *Australian Journal of Plant Physiology* **7**, 519–25.

Bechtel, D.B. (1985). The microstructure of wheat: its development and conversion into bread. *Food Microstructure* **4**, 125–33.

Bhullar, S.S. & Jenner, C.F. (1986). Effects of temperature on the conversion of sucrose to starch in the developing wheat endosperm. *Australian Journal of Plant Physiology* **13**, 605–15.

Briarty, L.G., Hughes, C.E. & Evers, A.D. (1979). The developing endosperm of wheat – a stereological analysis. *Annals of Botany* **44**, 641–58.

Brink, R.A. & Cooper, R.A. (1947). The endosperm in seed development. *Botanical Reviews* **13**, 423–541.

Caley, C.Y. (1986). *Termination of grain growth in cereals*. Ph.D. thesis, University of Edinburgh, Scotland.

Caley, C.Y., Duffus, C.M. & Jeffcoat, B. (1990). Photosynthesis in the pericarp of developing wheat grains. *Journal of Experimental Botany*, **41**, 303–7.

Cameron-Mills, V. & Duffus, C.M. (1977). The *in vitro* culture of immature barley embryos on different culture media. *Annals of Botany* **41**, 1117–27.

Cameron-Mills, V. & Duffus, C.M. (1979). Sucrose transport in isolated immature barley embryos. *Annals of Botany* **43**, 559–69.

Chojecki, A.J.S., Gale, M.D. & Bayliss, M.W. (1986). The number and sizes of starch granules in the wheat endosperm and their association with grain weight. *Annals of Botany* **58**, 819–31.

Cobb, B.G., Hole, D.J., Smith, J.D. & Kent, M.W. (1988). The effects of modifying sucrose concentration on the development of maize kernels grown *in vitro*. *Annals of Botany* **62**, 265–70.

Cochrane, M.P. (1983). Morphology of the crease region in relation to assimilate uptake and water loss during caryopsis development in barley and wheat. *Australian Journal of Plant Physiology* **10**, 473–91.

Cochrane, M.P. (1985). Assimilate supply and water loss in maturing barley grains. *Journal of Experimental Botany* **36**, 770–82.

Cochrane, M.P. & Duffus, C.M. (1979). Morphology and ultrastructure of immature cereal grains in relation to transport. *Annals of Botany* **44**, 67–72.

Cochrane, M.P. & Duffus, C.M. (1981). Endosperm cell number in barley. *Nature* **289**, 399–401.

Day, M.D., Bayliss, M.W. & Pryke, J.A. (1987). Plastid DNA during grain filling in wheat. *Plant Science* **53**, 131–8.

Doehlert, D.C., Kuo, T.M. & Felker, F.C. (1988). Enzymes of sucrose and hexose metabolism in developing kernels of two inbreds of maize. *Plant Physiology* **86**, 1013–19.

Donovan, G.R., Jenner, C.F., Lee, J.W. & Martin, P. (1983). Longitudinal transport of sucrose and amino acids in the wheat grain. *Australian Journal of Plant Physiology* **10**, 31–42.

Duffus, C.M. (1984). Metabolism of reserve starch. In *Storage Carbohydrates in Vascular Plants* (ed. D.H. Lewis) (Society for Experimental Biology Symposium vol. 19), pp. 231–52. Cambridge University Press.

Duffus, C.M. (1987). Physiological aspects of enzymes during grain development and germination. In *Enzymes and Their Role in Cereal Technology* (ed. J.E. Kruger, D. Lineback & C.E. Stauffer), pp. 83–164. St Paul, Minnesota: American Association of Cereal Chemists.

Duffus, C.M. (1990). Recent advances in the physiology and biochemistry of cereal grains in relation to pre-harvest sprouting. In *Proceedings of the Fifth International Symposium on Pre-Harvest Sprouting in Cereals* (ed. K. Ringlund, E. Mosleth & D.J. Mares), pp. 47–56. Boulder, Colorado: Westview Press.

Duffus, C.M. & Binnie, J. (1990). Sucrose relationships during endosperm and embryo development in wheat. *Plant Physiology and Biochemistry*, **28**, 161–5.

Duffus, C.M. & Cochrane, M.P. (1982). Carbohydrate metabolism during cereal grain development. In *The Physiology and Biochemistry of Seed Development, Dormancy and Germination* (ed. A.A. Khan), pp. 43–66. Amsterdam: Elsevier Biomedical Press.

Duffus, C.M. & Murdoch, S.M. (1979). Variation in starch granule size distribution and amylose content during wheat endosperm development. *Cereal Chemistry* **36**, 427–9.

Duffus, C.M., Nutbeam, A.R. & Scragg, P.A. (1985). Photosynthesis in the immature cereal pericarp in relation to grain growth. In *Regulation of Sources and Sinks in Crop Plants* (ed. B. Jeffcoat, A.F. Hawkins & A.D. Stead), pp. 243–56. Long Ashton, England: British Plant Growth Regulator Group.

Duffus, C.M. & Rosie, R. (1978). Metabolism of ammonium ion and glutamate in relation to nitrogen supply and utilisation during grain development in barley. *Plant Physiology* **61**, 570–4.

Duffus, C.M. & Slaughter, J.C. (1980). *Seeds and Their Uses*. Chichester: John Wiley & Sons.

Echeverria, E., Boyer, C.D., Thomas, P.A., Kang-Chien, L. & Shannon, J.C. (1988). Enzyme activities associated with maize endosperm amyloplasts. *Plant Physiology* **86**, 786–92.

Entwistle, G. & ap Rees, T. (1988). Enzymic capabilities of amyloplasts from wheat (*Triticum aestivum*) endosperm. *Biochemical Journal* **255**, 391–6.

Evans, L.T. & Rawson, H.M. (1970). Photosynthesis and respiration by the flag leaf and components of the ear during grain development in wheat. *Australian Journal of Biological Sciences* **23**, 245–54.

Fisher, D.B. & Gifford, R.M. (1986). Accumulation and conversion of sugars by developing wheat grains. *Plant Physiology* **82**, 1024–30.

Gougler Schmalstig, J. & Hitz, W.D. (1987). Transport and metabolism of a sucrose analogue (1'-fluorosucrose) into *Zea mays* L. endosperm without invertase hydrolysis. *Plant Physiology* **85**, 902–5.

Griffith, S.M., Jones, R.J. & Brenner, M.L. (1987*a*). *In vitro* sugar transport in *Zea mays* L. kernels. I. Characteristics of sugar absorption and metabolism by developing maize endosperm. *Plant Physiology* **84**, 467–71.

Griffith, S.M., Jones, R.J. & Brenner, M.L. (1987*b*). *In vitro* sugar transport in *Zea mays* L. kernels. II. Characteristics of sugar absorption and metabolism by isolated developing embryos. *Plant Physiology* **84**, 472–5.

Hayashi, H. & Chino, M. (1986). Collection of pure phloem sap from wheat and its chemical composition. *Plant Cell Physiology* **27**, 1387–93.

Ho, L.C. (1988) Metabolism and compartmentation of imported sugars in sink organs in relation to sink strength. *Annual Review of Plant Physiology and Plant Molecular Biology* **39**, 355–78.

Huber, A.G. & Grabe, D.F. (1987). Endosperm morphogenesis in wheat: termination of nuclear division. *Crop Science* **27**, 1252–6.

Jenner, C.F. & Rathjen, A.J. (1972). Factors limiting the supply of sucrose to the developing wheat grain. *Annals of Botany* **36**, 729–41.

Jenner, C.F. & Rathjen, A.J. (1977). Supply of sucrose and its metabolism in developing grains of wheat. *Australian Journal of Plant Physiology* **4**, 691–701.

Keeling, P., Wood, J.R., Tyson, R.H. & Bridges, I.C. (1988). Starch biosynthesis in developing wheat grain. Evidence against the direct involvement of triose phosphates in the metabolic pathway. *Plant Physiology* **87**, 311–19.

Kent, N.L. (1983). *Technology of Cereals*. Oxford: Pergamon Press.

Kvaale, A. & Olsen, O.A. (1986). Rates of cell division in developing barley endosperms. *Annals of Botany* **57**, 829–33.

Lemoine, R., Delrot, S., Gallet, O.J. & Larsson, C. (1989). The sucrose carrier of the plant plasma membrane. II. Immunological characterisation. *Biochimica et Biophysica Acta* **978**, 65–71.

Lingle, S.E. & Chevalier, P. (1984). Movement and metabolism of sucrose in developing barley kernels. *Crop Science* **24**, 315–19.

Lyznik, L. & Tsai, C.Y. (1989). Protein synthesis in endosperm cell culture of maize (*Zea mays* L.). *Plant Science* **63**, 105–14.

MacLeod, L.C. & Duffus, C.M. (1988a). Reduced starch content and sucrose synthase activity in developing endosperm of barley plants grown at elevated temperatures. *Australian Journal of Plant Physiology* **15**, 367–75.

MacLeod, L.C. & Duffus, C.M. (1988b). Temperature effects on starch granules in developing barley grains. *Journal of Cereal Science* **8**, 29–37.

Morrison, I.N., Kuo, J. & O'Brien, T.P. (1975). Histochemistry and fine structure of developing wheat aleurone cells. *Planta* **123**, 105–16.

Morrison, W.R. (1988). Lipids in cereal starches: A review. *Journal of Cereal Science* **8**, 1–15.

Nicholas, M.E., Gleadow, R.M. & Dalling, M.J. (1984). Effects of drought and high temperature on grain growth in wheat. *Australian Journal of Plant Physiology* **11**, 553–66.

Nutbeam, A.R. & Duffus, C.M. (1978). Oxygen exchange in the pericarp green layer of immature cereal grains. *Plant Physiology* **62**, 360–2.

Ouattar, S., Jones, R.J., Crookston, R.K. & Kajeiou, M. (1987). Effect of drought on water relations of developing maize kernels. *Crop Science* **27**, 730–5.

Palmer, G.H. (1989). Cereals in malting and brewing. In *Cereal Science and Technology* (ed. G.H. Palmer), pp. 61–242. Aberdeen University Press.

Parker, M.L. (1985). The relationship between A-type and B-type starch granules in the developing endosperm of wheat. *Journal of Cereal Science* **3**, 271–8.

Peterson, D.M., Saigo, R.H. & Holy, J. (1985). Development of oat aleurone cells and their protein bodies. *Cereal Chemistry* **62**, 366–71.

Porter, G.A., Knievel, D.P. & Shannon, J.C. (1985). Sugar efflux from maize (*Zea mays* L.) pedicel tissue. *Plant Physiology* **77**, 524–31.

Porter, G.A., Knievel, D.P. & Shannon, J.C. (1987). Assimilate unloading from maize (*Zea mays* L.) pedicel tissues. *Plant Physiology* **83**, 131–6.

Preiss, J. (1988). Biosynthesis of starch and its regulation. In *The Biochemistry of Plants*, vol. 14 (ed. J. Preiss), pp. 181–254. New York: Academic Press.

Rahman, S., Kreis, M., Forde, B.G., Shewry, P.R. & Miflin, B.J. (1984). Hordein gene expression during development of the barley (*Hordeum vulgare*) endosperm. *Biochemical Journal* **223**, 315–22.

Reeves, C.D., Krishnan, H.B. & Okita, T.W. (1986). Gene expression in developing wheat endosperm: accumulation of gliadin and ADP glucose pyrophosphorylase messenger RNAs and polypeptides. *Plant Physiology* **82**, 34–40.

Rijven, A.H.G.C. (1986). Heat inactivation of starch synthase in wheat endosperm. *Plant Physiology* **81**, 448–53.

Smart, M.G. & O'Brien, T.P. (1983). The development of the wheat embryo in relation to the neighbouring tissues. *Protoplasma* **114**, 1–13.

Tashiro, T. & Wardlaw, I.F. (1989). A comparison of the effect of high temperature on grain development in wheat and rice. *Annals of Botany* **64**, 59–65.

Tercé-Laforgue, T., Sallantin, M. & Pernollet, J.-C. (1987). Wheat endosperm in m-RNA and polysomes and their *in vitro* translation products during development and early stages of germination. *Physiologia Plantarum* **69**, 105–12.

Vieweg, G.H. & de Fekete, M.A.R. (1980). On the effects of lipids on starch-metabolizing enzymes and its significance in relation to the simultaneous synthesis of amylose and amylopectin in starch granules. In *Mechanisms of Saccharide Polymerization and Depolymerization* (ed. J.J. Marshall), pp. 175–85. New York: Academic Press.

Watson, P.A. & Duffus, C.M. (1988). Carbon dioxide fixation by detached cereal caryopses. *Plant Physiology* **87**, 504–9.

Watson, P.A. & Duffus, C.M. (1991). Light-dependent CO_2 retrieval in mature barley caryopses. *Journal of Experimental Botany* **42**, 1013–19.

Weber, E.J. (1969). Lipids of maturing grain of corn (*Zea mays* L.). *Journal of the American Oil Chemists' Society* **46**, 485–8.

Wolswinkel, P. (1988). Nutrient transport into developing seeds. In *ISI Atlas of Science, Animal and Plant Sciences*, vol. 1, pp. 298–302.

Wolswinkel, P. & Amerlaan, A. (1989). Effect of the osmotic environment on assimilate transport in isolated developing embryos of maize (*Zea mays* L.). *Annals of Botany* **63**, 705–8.

A. G. STEPHENSON

The regulation of maternal investment in plants

Introduction

The effects of the environment on the number and quality of the offspring produced by organisms is *the* fundamental concern of evolutionary ecology. For plants, the number and quality of their offspring may be limited by external environmental factors, such as pollination and seed predation, or internally by the availability of resources which, in turn, may be influenced by weather and soil conditions, herbivory, disease and competition. Each of these factors may vary over time and location. In response to these varying environmental conditions, plants may adjust their resource commitment to seed production by altering (i) the number (or gender) of flowers that differentiate, (ii) the number of ovules per flower, (iii) the number of fruits and seeds that abort, and (iv) the mass of the mature seeds (see, for example, Adams, 1967; Primack, 1978; Lloyd, 1980; Stephenson, 1981, 1984; Thomson, 1989). In the parlance of our colleagues who study animals, plants can potentially alter (i) clutch size (a clutch is defined as offspring in more or less the same developmental phase that are simultaneously drawing resources from the maternal plant), (ii) the interval between successive clutches, and/or (iii) the mass of the offspring at the termination of maternal care (Stearns, 1976; Burley, 1980).

The goals of this Chapter are to review the individual mechanisms by which plants regulate maternal investment under various environmental conditions, to review the directions in which these mechanisms are regulated with respect to one another, and to review the consequences of the different mechanisms of regulation on offspring performance (quality). I attempt to achieve these goals in a rather unorthodox manner. Firstly, a series of experiments is described using the common zucchini (*Cucurbita pepo* cv. Black Beauty Bush, Cucurbitaceae) and then I attempt to achieve the requisites of a review via a series of discussions following each experiment. It should be noted that, for an outcrossing hermaphroditic plant, its seed crop represents only half of its nuclear genetic contribution to the next generation and, consequently, by focusing on the regulation of maternal investment this Chapter focuses on only half of the reproductive story (see Bertin, 1988).

Experiment I. Effects of fruit production

The Black Beauty Bush cultivar of zucchini is a monoecious, annual, short-internode vine with indeterminate growth and reproduction. The large, yellow flowers are borne individually and are bee-pollinated; each remains open for only a few hours on one morning. After an initial period of vegetative growth, one pistillate flower differentiates at each node, near the apex of the plant. Later, one (occasionally two) staminate flower is formed at the older nodes. A typical ovary contains 800 or more ovules of which only 300 or so produce mature seeds. Each viable seed weighs 0.10 – 0.24 g. The mature fruit is cylindric in shape and has a volume of approximately three litres (Winsor, Davis & Stephenson, 1987; Stephenson, Devlin & Horton, 1988a).

In natural populations of plants, flowers may fail to set fruit because of lack of pollination or because of damage or removal of flowers and immature fruits by herbivores. Experiment I was designed to compare the effects of flower and immature fruit loss to the effects of fruit maturation on the number of flowers produced by zucchini plants. Zucchini plants were sown at 2 m intervals in a 30 m × 100 m field at the Pennsylvania State University Agriculture Experiment Station near State College, Pennsylvania, USA. On 10 plants, every pistillate flower was removed on the day of anthesis throughout the growing season (pistillate flower removal treatment) (Stephenson et al., 1988a). On another 10 plants, every fruit was removed 5 d after anthesis (harvest treatment). Typically, these fruits were 15–22 cm in length at the time of harvest. On a third group of 15 plants, only 14 of which survived the entire season, the flowers were allowed to be open (bee) pollinated and the fruits were allowed to develop to maturity (controls). On these three treatments, staminate and pistillate flower production and fruit maturation or abortion (controls only) were recorded daily. A fruit was judged as aborted when it began to discolour and became noticeably soft. The remaining plants in the field were used for a series of non-destructive and destructive tests to determine the rate of fruit growth.

Results: control plants

Fruit growth was analysed by using non-linear regression and it was found that zucchini fruits, like those of most other species (see, for example, Bollard, 1970), have a sigmoidal growth curve when volume is plotted against age (Fig. 1). Moreover, we found that volume was highly cor-related ($R^2=0.79$) with dry mass. The initial period of slow growth occurred from 1 to 5 d post-anthesis, the period of rapid growth in which most of the resources are accumulated by the fruit was from 6 to 15 d post-anthesis, and the final period of slow growth extended from 16 to 20 d post-anthesis. The 14 control plants produced 9.9 ± 1.9 ($\bar{x}$ ± SD) pistillate flowers, of which 4.5 ± 1.1 produced mature fruit containing 340 ± 17 seeds. Fruit abortion occurred 8.4 ± 1.3 d after anthesis, but aborted fruits were considerably smaller at this time than were fruits that continued to develop. Log–linear models revealed that a pistillate flower was significantly less likely to produce a mature fruit (more likely to abort) when a 6–15 d old fruit was already on the vine. Moreover, on a daily basis, a plant was less likely to produce a pistillate flower when a 6–15 d old fruit was on the vine, indicating that some pistillate flowers were aborted prior to anthesis. This combination of flower and fruit abortion tended to create distinct intervals between successive fruits.

From this brief examination of the data from the control plants, it is possible to conclude that zucchini produces more pistillate flowers than it can carry to maturity, that the extra fruits are aborted before the period

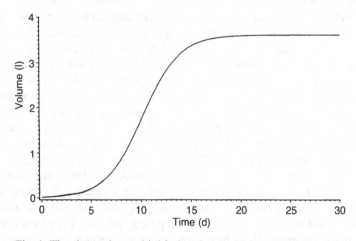

Fig. 1. The change in zucchini fruit volume with age. Time 0 is the day of anthesis. The curve was determined by the non-linear regression pro-cedure of SAS (1982).

of rapid resource accumulation, and that the order of flower production and the growth phase of the already developing fruits determine, to a large extent, which fruits mature and which abort.

Discussion: response of control plants

With few exceptions, outcrossing plants regularly produce far more flowers than mature fruits (see Lloyd, 1980; Stephenson, 1981; Sutherland & Delph, 1984; Lee, 1988). Hand-pollination studies, resource enrichment and deprivation studies, as well as other types of evidence, indicate that these species are physiologically incapable of providing the necessary resources to develop mature fruits from all of their flowers (Lloyd, 1980; Stephenson, 1981; Lee, 1988). That is, an upper limit to fruit production is set by resources rather than flower number. When the number of pollinated flowers exceeds the resources available for fruit maturation (as commonly occurs), immature fruits abort until fruit production matches resource availability (Lloyd, 1980; Stephenson, 1981; Lee, 1988). For many of these species, the presence of developing fruits (determined by the order of pollination and position of the flower) leads to higher abortion rates for later (younger) developing fruits and flowers (Stephenson, 1981; Lee, 1988). This pattern of maturation and abortion is often manifested over the entire plant in species with indeterminate flowering, in which individual flowers or inflorescences are produced sequentially during the growing season. This pattern can also be seen within inflorescences and is particularly apparent in many species with terminal inflorescences that develop acropetally. In these species, the first flowers to open enjoy not only a temporal but also a spatial advantage in that they are more favourably located in terms of access to the resources necessary for maturation (Stephenson, 1981; Lee, 1988). The physiological causes or chain of events leading to the abortion of flowers and younger fruits (often termed first-fruit dominance) are not clear and are beyond the scope of this Chapter. The hypotheses, however, range from simple competition among flowers and fruits (sinks) for resources, through the production of plant growth hormones that regulate source–sink interactions and effectively divert resources away from later sinks, to the production of growth regulators by the developing fruit that actively inhibit later (younger) reproductive structures (Bangerth, 1989). While the physiological chain of events that underlies first-fruit dominance is still subject to active debate, it is apparent that most fruit abortion on virtually all species that have been examined occurs early in development, before the period when most of the resources are accumulated in the fruit (Stephenson, 1981; Bawa & Webb, 1984).

Table 1. *The effect of pistillate flower removal and fruit removal treatments on pistillate and staminate flower production in zucchini*

(Values are means ± SD.)

Treatment	Number of plants	Number of pistillate flowers per plant[a]	Number of staminate flowers per plant[a]
All pistillate flowers removed at anthesis	10	28.6 ± 5.8a	26.4 ± 5.3a
Fruits harvested when 5 d old	10	24.3 ± 6.3a	30.9 ± 11.9a
Fruits allowed to develop to maturity	14	9.9 ± 1.9b	19.3 ± 4.3b

[a]Means followed by different letters differ significantly with a Bonferroni multiple pairwise comparison test (overall $\alpha = 0.05$).

Results: pistillate flower removal and harvest treatments

A comparison of the data obtained from the pistillate flower removal, the harvest and the control treatments revealed that the flower removal and harvest treatments did not significantly differ in either the number of pistillate (Stephenson *et al.*, 1988a) or the number of staminate flowers that they produced, but that both treatments produced significantly more staminate and pistillate flowers than the control treatment (Table 1).

Three conclusions concerning the regulation of maternal investment can be derived from this simple experiment. First, fruit and seed maturation depresses flower production but the harvest of five-day-old fruits does not. Secondly, there is a trade-off between the female function (fruit and seed production) and the male function (in this case, the number of staminate flowers) of zucchini. Thirdly, because staminate flower production is positively correlated with the number of nodes on zucchini, fruit production depresses vegetative growth.

Discussion: treatment responses

Many studies have also reported that a lack of pollination, the removal of flower buds, or the loss of young fruits increases subsequent flower production (Lloyd, 1980; Stephenson, 1981; Lovett Doust & Eaton, 1982; Marshall, Levin & Fowler 1985; Stanton, Bereczky & Hasbrouck, 1987). For example, in a greenhouse study, Stanton *et al.* (1987) found that plants of *Raphanus raphanistrum* that were unpollinated produced 210% more flowers than plants in which every flower was hand-pollinated. Moreover, this increase in flower production closely paralleled a 192% increase in the production of lateral meristems. In addition, intermediate levels of pollination produced intermediate levels of flower production. The negative effects of fruit production on both vegetative growth and flower production can also be clearly seen in many commercial fruit and nut species that have a tendency for biennial flowering and fruiting cycles (Stephenson, 1981; Goldschmidt & Golomb, 1981; Sparks & Madden, 1985; Schaffer *et al.*, 1985). In these species, the size of the current fruit crop is negatively correlated with vegetative growth. In years with small (or no) fruit crops, many flower buds differentiate at the end of the growing season and open the following spring. In years with large fruit crops, few buds differentiate. Horticulturalists can eliminate periodicities in the fruiting cycle by artificially thinning the flowers and young fruits which, in turn, increases growth, stored carbohydrate and nutrients, and flower-bud production later in the season. Moreover, the thinning of fruits from some branches and not others leads to branches that bear annually and others that bear biennially, indicating that shoots are semi-autonomous (Harper & White, 1974; Watson & Casper, 1984). The effects of thinning, however, are strongly dependent upon the age of the fruits. The loss of fruits after their rapid growth phase has little effect on vegetative growth or flower production (Stephenson, 1981; Marshall *et al.*, 1985).

On some important tree species, the loss of flowers and immature fruits due to flower and seed predation or inclement weather also leads to increases in stored resources, growth and flower primordia in a manner similar to artificial thinning (see Stephenson, 1981). For example, in coconuts (*Cocos nucifera*), damage by a seed predator (*Pseudotheraptus wayi*) leads to the abscission of young fruits and a corresponding increase in the subsequent production of flower primordia (Vanderplank, 1958, 1960).

For species with perfect flowers, the negative relationship between fruit maturation and flower production directly affects a plant's potential for reproduction through the male function (Lloyd, 1980; Stephenson, 1981; Stanton *et al.*, 1987). In contrast, the abortion of young fruits on plants with perfect flowers regulates maternal investment in a manner that does not affect the opportunities for pollen dissemination (Lloyd, 1980; Stephenson, 1984). In this regard, Sutherland & Delph (1984) surveyed the literature and compared the proportion of ovary-bearing flowers that produced mature fruit on perfect-flowered inbreeders (where the production of additional pollen is of little advantage), monoecious species (where pistillate flower production can be adjusted independently of pollen production) and perfect-flowered outcrossers. They found that perfect-flowered inbreeders abort very few fruits while perfect-flowered outcrossers abort the highest percentage of fruits. This indicates that the male function of a plant can influence, at least to some degree, whether the regulation of maternal investment occurs via flower differentiation or via the abortion of immature fruits. In sequential hermaphrodites and andromonoecious plants, the gender of the flowers is often regulated in response to prior fruit production (Solomon, 1986; Schlessman, 1988). For example, in the andromonoecious plant *Solanum carolinense*, the proportion of staminate flowers within an inflorescence changes with fruit production and the proportion of staminate flowers changes between inflorescences with the level of fruit production on the earlier inflorescences (Solomon, 1986).

Experiment II. Effects of resource enrichment or deprivation

In natural plant populations, rainfall, soil fertility, herbivory, competition and other factors can alter the amount of resources available for reproduction. Consequently, in the second experiment, we (T.-C. Lau & A.G. Stephenson) examined the effects of soil nitrogen levels on reproduction in zucchini. Seventy-five zucchini seeds were planted at 2 m intervals at the field site in a field that exceeded the recommended levels of soil phosphorus and potassium for commercial *C. pepo* production but was very deficient in nitrogen. Three nitrogen treatments (low, medium and high), each consisting of 25 plants, were created by adding slow-release nitrogen pellets to the soil around each plant. Throughout the growing season staminate and pistillate flower production and fruit maturation or abortion were monitored, and in the autumn the mature fruits were harvested and seed number and seed mass were determined.

Results

The analyses of the resulting data revealed that the treatments differed significantly in the number of days from emergence to the production of the first pistillate flower (about six days earlier in the high N treatment), the number of both staminate and pistillate flowers, the number of mature fruit (high N = 3.7 ± 1.1 fruits per plant; low N = 2.8 ± 1.3 fruits per plant; $\bar{x} \pm$ SD), and total seed number and seed mass per plant. There were no significant differences in seed number per fruit or mean seed mass (although the high N treatment exceeded the low and medium treatments in each case) or in the proportion of flowers that produced mature fruit (about 55% in each treatment). When the mature fruits were divided into those initiated during first (July) and second (August) halves of the growing season, both seed number per fruit and mean seed weight were significantly decreased during the second half of the growing season, but there was no season by N-treatment interaction, indicating that the decreases in seed number per fruit and mean seed mass were similar for the three N treatments.

From these data we conclude that, under the nitrogen conditions in this experiment, zucchini regulates maternal investment merely by decreasing the length of the period of vegetative growth and by slightly decreasing the interval between successive pistillate flowers and fruits during the period of reproductive (and vegetative) growth. Across all three N treatments, there is a negative relation between previous fruit and seed production and seed number per fruit and mean seed mass per fruit.

Discussion

In other species, the effects of nutrient enrichment or deprivation have been reported to cover the entire range of mechanisms for regulating maternal investment (Primack 1978; Willson & Price, 1980; Stephenson, 1981, 1984; Schlichting & Levin, 1984; Marshall, Levin & Fowler, 1986; Lee, 1988). Often the effects of nutrient enrichment or deprivation depend on the time of application. For example, in many commercial fruit trees, the application of fertilizer during an off (vegetative) year or late in the season in a reproductive year will increase the number of flower buds that differentiate. However, if the fertilizer is added in the spring of a reproductive year, there is an increase in the proportion of flowers that produce fruit and an increase in mean seed mass (see Stephenson, 1981).

Many studies have also reported a decrease in seed number and/or mean seed mass per fruit in the fruits produced later in the growing season (Stephenson, 1981; Lee, 1988; see also Marshall *et al.*, 1985, 1986; Galen, Plowright & Thomson, 1985; Stanton *et al.*, 1987; Devlin, 1989). For example, in *Lobelia cardinalis*, which produces terminal inflorescences that develop acropetally, seed number and mean seed weight varied systematically with floral position on the inflorescence. The flowers on the top (later) one-third of the inflorescence produced significantly fewer and smaller seeds than the flowers on the bottom one-third of the inflorescence (Devlin, 1989). Often such changes in seed number and/or mean seed mass per fruit are reversible if fruit set is prevented on the earlier flowers (see, for example, van Steveninck, 1957; Marshall *et al.*, 1985; Galen *et al.*, 1985; Stanton *et al.*, 1987) indicating that prior fruit and seed production has a negative effect on the later fruits, probably by depleting the resources needed for fruit and seed maturation.

These changes in seed number per fruit could be due to either a decrease in the number of ovules per flower and/or to a decrease in the number of seeds that mature (increase in seed abortion). Thomson (1989) has recently reviewed the scanty literature on variations in ovule number per flower within individuals. He concludes that ovule number per flower commonly declines in successively opening flowers within the inflorescence. Moreover, Stanton *et al.* (1987) showed that mean ovule number per flower in *Raphanus raphanistrum* decreases over the growing season on pollinated plants but does not decrease on unpollinated plants, indicating that depletion of resources by early fruits reduces ovule number on the later flowers (see also Vasek *et al.*, 1987).

The relation between ovule number and seed number, however, is not well understood. Most outcrossing plants produce far more ovules than mature seeds even when the flowers are hand-pollinated (see Wiens, 1984; Lee, 1988). Direct observation and other methods have revealed that most species commonly abort seeds (Lee, 1988). The abortion of some seeds is undoubtedly due to the expression of lethal recessives in the embryo or endosperm. For example, self-pollination on a normally outcrossing plant leads to greater levels of seed abortion than self-pollination on a normally inbreeding plant, which has presumably been purged of its lethal recessives in earlier generations (e.g., Levin, 1984). On many species, however, seed abortion has been shown to increase systematically as the number of developing fruits increases and, in some species, seed abortion has been shown to decrease if the earlier flowers are prevented from producing fruit (Lee, 1988).

The effects of position and time of initiation also extend to seeds within fruits. In species with linearly arranged ovules, the position of the ovule often has a strong effect on the probability of maturation and the weight of the mature seed (Lee, 1988; Nakamura, 1988; Rocha & Stephenson, 1990) indicating that some ovule positions are at a temporal and spatial disadvantage in garnering resources. In *Phaseolus coccineus*, the probability of maturation and the size of the seeds in the disadvantaged ovule positions increases when the ovules in the favoured positions are destroyed within 48 hours of fertilization (O.J. Rocha & A.G. Stephenson, unpublished data). Nakamura (1988) has been able to successfully mature cultured embryos from the disadvantaged ovule positions in *Phaseolus vulgaris*. Clearly, viable seeds are aborted in these species.

The effects of prior fruit and seed production may or may not have a corresponding effect on the male function of a plant. For example, in *Lobelia cardinalis*, pollen production per flower does not change with the position of the flower or the number of developing fruits on the inflorescence (Devlin, 1989). In zucchini, however, pollen production per staminate flower does decrease significantly through the growing season (Small, 1988). Vasek *et al.* (1987) showed that pollen production per flower in *Clarkia unguiculata* increased with the addition of nutrients and decreased with the number of developing fruits (nodal position).

Experiment III. Variations in pollen deposition and seed number

In natural populations of plants, the intensity of pollination can vary both within and between plants. Consequently, in the third experiment, the number of pollen grains that were deposited onto the stigmas was systematically varied in 60 plants that were growing at 2 m intervals at the field site. On 15 plants every pistillate flower received a low pollen load (240 ± 36 grains, $\bar{x}$ ± SD). On another 15 plants, all pistillate flowers received a medium pollen load (2 × low load). On a third group of 15 plants, all pistillate flowers received a large pollen load in which the stigmas were saturated with pollen (> 5000 grains). On the final 15 plants, low, medium and high pollen loads were alternated on the successive pistillate flowers produced by each plant. On five of these plants, the first flower received a low load, the second a medium load and the third a high pollen load. The cycle was then repeated for the additional flowers. On another five plants the cycle began with a medium pollen load and on the final five plants the cycle began with a high pollen load. These four groups of 15 plants are termed the low, medium, high and alternate plants. Throughout the growing season the flower and fruit production of each plant was recorded. In the autumn the mature fruits were harvested and their seeds counted and weighed (Winsor *et al.*, 1987; Stephenson *et al.*, 1988a).

Results

For the low, medium and high pollen-load plants, analyses of variance revealed that the size of the pollen load had a significant effect on seed number per fruit (low pollen load = 39.5 ± 3.2 seeds per fruit; medium =

55.9 ± 4.9; high = 375 ± 17.5; $\bar{x}$ ± SE) and mean seed mass per fruit (low = 0.190 ± 0.003 g; medium = 0.189 ± 0.003 g; high = 0.138 ± 0.003 g; $\bar{x}$ ± SE) (Davis, 1986; Winsor *et al.*, 1987). In addition, the low and medium pollen-load plants produced significantly more pistillate flowers (low = 16.2 ± 2.6; medium = 15.4 ± 4.2; high = 9.3 ± 1.5; $\bar{x}$ ± SE) and mature fruits (low = 8.1 ± 1.9; medium = 7.5 ± 1.7; high = 4.4 ± 1.0; x ± SE) than did the high pollen-load plants. However, a χ^2 test revealed that the probability that a pistillate flower will produce a mature fruit is independent of the size of the pollen load ($\chi^2 = 0.31$; df = 2; $p > 0.35$). In short, when the size of the pollen load is varied *between* plants, about 50% of the pistillate flowers produce a mature fruit with all three pollen load treatments.

Log–linear models revealed that the presence of one or more 6–15 d old fruits (the period of greatest resource accumulation) decreased the probability that a pistillate flower will produce a mature fruit (increases the probability of abortion). However, the strength of this first fruit dominance was much stronger on the high pollen-load plants than on the low and medium pollen-load plants. For example, one 6–15 d old fruit on the high pollen-load plants decreased the probability that a pistillate flower will produce a mature fruit more than the presence of one 6–15 d old fruit on the low and medium pollen-load plants (Stephenson *et al.*, 1988*a*).

On the alternate pollen-load plants, the three sizes of pollen loads also produced significant differences in seed number per fruit (low = 13 ± 7.8; medium = 77.4 ± 30.7 and high = 289.6 ± 26.5; $\bar{x}$ ± SE) and mean seed mass (low = 0.180 ± 0.004 g; medium = 0.176 ± 0.004 g; high = 0.156 ± 0.004 g; $\bar{x}$ ± SE). In contrast to the low, medium and high pollen-load plants, when the size of the pollen load was varied among flowers *within* a plant (the alternate plants), the probability that a flower will produce a mature fruit was not independent of the size of the pollen load ($\chi^2 = 40.6$; df = 2; $p < 0.001$). The flowers receiving high pollen loads were far more likely to produce a mature fruit (71%) than were the flowers receiving medium (28%) or low (18%) pollen loads (Winsor *et al.*, 1987). That is, young fruits with high seed numbers can often override the dominance of older fruits with lower seed numbers.

From this experiment it is reasonable to conclude that zucchini responds to pollination levels that are inadequate to produce a full complement of seeds by increasing seed size and by altering the interval between successive fruits when the variations in seed number occur *between* plants, and by the non-random abortion of fruits on the basis of seed number when the variation in seed number occurs *within* plants.

Discussion

A comparison of the N addition experiment and the alternate pollen-load treatment reveals that the direction of the relation between seed number per fruit and mean seed mass per fruit can change with the experimental conditions. In the N addition experiment, seed number per fruit and mean seed mass were positively correlated: both decreased during the second half of the growing season. In contrast, in the alternate pollen-load treatment of this experiment, seed number per fruit is negatively correlated with mean seed mass: the seeds in fruits resulting from low and medium pollen loads were significantly heavier than those from fruits resulting from high pollen loads. It should be noted, however, that the seeds resulting from low and medium pollen loads in the alternate treatment weighed significantly less than the seeds resulting from the low and medium pollen-load treatments (where all flowers on a plant received the same size of pollen load) whereas the seeds resulting from high pollen loads on the alternate treatment weighed significantly more than the seeds resulting from the high pollen load treatment. These findings indicate that seed mass is a function of the number of seeds vying for maternal resources within a fruit and the total resources available, which, in turn, is influenced by the number of seeds in the other fruits on a plant.

The number of descendants that a plant leaves through the female function is not only determined by the number of seeds it produces but also by the vigour of its seeds. Numerous studies have revealed that seed mass affects progeny vigour, especially under competitive conditions (see, for example, Black, 1956; Schaal, 1980; Gross, 1984; Stanton, 1984; Mazer, 1987). In general, larger seeds are more likely to germinate and survive, and they tend to grow faster and have greater reproductive output as adults, but there are exceptions. Studies of progeny vigour that we have conducted using the zucchini seeds produced by the low and high pollen-load plants also reveal that seed size has a significant effect on progeny performance (Winsor et al., 1987). That is, large seeds from the low pollen-load plants outperform smaller seeds from the same plants and large seeds from the high pollen-load plants outperform smaller seeds from high pollen-load plants. However, seeds from the high pollen-load plants outperform comparably sized seeds from the low pollen-load plants. Moreover, average-sized seeds from the high-load plants significantly outperform average-sized seeds from the low-load plants despite a 27% difference in seed mass (Stephenson et al., 1988b), indicating that the size of the pollen load (or the number of seeds per fruit) has some beneficial effect on progeny vigour that is independent of seed size. Although the cause of this and similar pollen load and seed number effects on other species has been the subject of much debate and active investigation (see Stephenson *et al.*, 1988*b*; Charlesworth, 1988; Snow & Mazer, 1988; Schlichting *et al.*, 1990) it is reasonable to assume that the differences in the vigour of the progeny from low and high pollen-load plants would be even greater if the low pollen-load plants did not produce significantly larger seeds than the high pollen-load plants.

As with the naturally pollinated plants in Experiment I, the high pollen-load plants tended to have distinct cycles of fruit production and flower and fruit abortion that were associated with the presence or absence of 6–15 d old fruit. On the low and medium pollen-load plants, the order of pollination and the age of the developing fruits also have a significant effect on the probability that a flower will produce a mature fruit. However, the strength of the dominance exerted by any single fruit is diminished on the low and medium pollen-load plants compared with the high-load plants. This tends to erase the distinct cycles of fruit maturation–abortion by decreasing the interval between successive fruits, and leads to a progression of fruits that overlap in their development.

Analogous events occur in some bird species that produce multiple clutches during the summer. Decreases in clutch size (due to predation or egg removal manipulations) decrease the interval between successive clutches (see Burley, 1980). In plants, similar events can occur on trees that tend to bear fruit biennially. As discussed earlier, the same trees will bear annually (and produce heavier seeds) if the fruit crop is artificially thinned or if some fruits are damaged. On plants of *Lotus corniculatus*, a herbaceous perennial that produces an indeterminate number of inflorescences over the growing season, the addition of fertilizer decreased the interval between successive inflorescences (increased the number of inflorescences) and increased the mass of the seeds, but did not alter the proportion of flowers that produced mature fruit compared with control plants (Stephenson, 1984).

Several studies have shown that plants selectively abort fruits on the basis of seed number (Stephenson, 1981; Lee, 1988). In all cases, those fruits with the fewest seeds are the most likely to abort. However, few studies have examined both the effects of first-fruit dominance and seed number on fruit maturation. Lee & Bazzaz (1982*a,b*) showed that the order of pollination influences the probability of fruit maturation on *Cassia fasciculata* but they also found that deviations from this pattern are due to a tendency to mature those fruits with the most seeds. Similarly, Bertin (1982) found that the order of pollination has a strong effect on the probability of fruit maturation on *Campsis radicans* but that deviations from this pattern are associated with below average levels of pollen deposition.

On the alternate plants in the zucchini experiment, most of the variation in seed number was due to variations in the size of the pollen loads. In natural plant populations, variations in seed number per fruit can be related to ovule number and seed abortion as well as to variations in the size of the pollen load. These, in turn, can be related to resource availability and to the congruence of the micro- and megagametophytic genomes (Lee, 1988). For many species, the abortion of fruits on the basis of seed number is resource efficient in that it costs less per seed to package many seeds in a fruit. For example, in zucchini a regression of dry mass of the pericarp per seed on the number of seeds per fruit, using a sample of 60 fruits with seed numbers ranging from 40 to 529, is highly significant ($R^2 = -0.59$; $p < 0.01$). This economy of seed packaging has also been identified in *Cercidium floridum* (Mitchell, 1977), *Asclepias speciosa* (Bookman, 1984), *Cassia fasciculata* (Lee & Bazzaz, 1982*b*), and other species (see Lee, 1988). Consequently, when seed number per fruit varies within a plant, the amount of resources saved by the more efficient packaging of seeds is sufficient to defray much of the cost of producing extra flowers (flowers above the number of fruits that can be carried to maturity) and then aborting the immature fruits.

The abortion of fruits on the basis of seed number may also have a beneficial effect on the average performance of the offspring. As we have noted for zucchini, seeds from the high pollen-load plants outperform comparably sized seeds from the low-load plants (Winsor *et al.*, 1987). Several other studies have found a similar relationship between size of the pollen load or seed number per fruit and progeny performance (see Lee, 1988) even when mean seed mass does not differ between the many- and fewer-seeded fruits (see, for example, Stephenson & Winsor, 1986). (The several hypotheses that have been advanced to explain this relationship have recently been reviewed by Schlichting *et al.* (1990).) Consequently, by aborting fruits on the basis of seed number, a plant may actually improve the average performance of the offspring it produces.

Summary and conclusions

In order to identify the effects of the environment on the number and quality of the offspring produced by animal species, evolutionary ecologists have given much attention to intraspecific variations in clutch size, number of clutches per breeding season, patterns of brood reduction and the mass of the offspring at the termination of parental care (e.g., Lack, 1954, 1966; Howe, 1976; Low, 1978; Burley, 1980). Despite difficulties in obtaining large sample sizes, in manipulating the relevant variables, and in quantifying important environmental conditions, such as food availability, these studies have provided cornerstones in current life-history theory (Cody, 1966; Giesel, 1976; Stearns, 1976). Plants, however, are more amenable to experimental manipulation and, therefore, may ultimately provide the necessary tests of the theory.

If we consider the seeds within a fruit to be a clutch (they are, after all, a group of offspring in more or less the same developmental phase in which the maternal plant is investing resources), then the evidence presented here indicates that zucchini can adjust clutch size, the interval between successive clutches, and the weight of the offspring at the termination of parental care. Clutch size was adjusted downward during the second half of the growing season in the N addition experiment by a decrease in ovule number, an increase in seed abortion, or some combination of the two. This appears to be a response to declining resource availability due, presumably, to prior fruit and seed production. Clutch size was also adjusted in the alternate pollen-load treatment via the selective abortion of fruits on the basis of seed number. Zucchini, as with many species, produces more pistillate flowers than it can develop to maturity and the non-random elimination of small clutches increases the average size of the remaining clutches.

The interval between successive clutches was adjusted in the low, medium, and high pollen-load treatments by altering the number of pistillate flowers that were produced and by the abortion of flowers and young fruits. When clutch size was large (the high pollen-load treatment) only one clutch (fruit) at a time tended to pass through the period of development requiring the greatest resource demand on the maternal parent. In short, there was little overlap in the development of successive clutches. In contrast, when clutch size was small (low pollen-load treatment) successive clutches tended to overlap in their development. Finally, zucchini regulated maternal investment by adjusting seed size downwards during the second half of the growing season in the N addition experiment when resources were presumably scarce owing to prior fruit and seed production, and by adjusting seed size upwards in the low and medium pollen-load treatments when resource availability exceeded the demand of the small clutches.

The zucchini data also indicate that clutch size, clutch interval and seed weight are complementary rather than alternative means of regulating maternal investment. For example, in the low, medium and high pollen-load treatments, both clutch interval and seed size changed in response to the variations in clutch size. In the N addition experiment, both clutch size and seed mass decreased as the amount of resources invested in prior fruit and seed production increased. In addition, plants often possess alternative mechanisms for adjusting clutch size and the interval between successive clutches. For example, clutch interval can be adjusted by altering the number of flower buds that differentiate, by altering the proportion of flowers and young fruits that abort, or by some combination of the two.

From a developmental and physiological perspective, the regulation of maternal investment can be viewed as merely the logical outcome of the rules governing competition for limited resources among the various vegetative and reproductive sinks. Weak sinks fail when resources are scarce or when many sinks are present. Unfortunately, our understanding of the rules determining sink strength is incomplete but the data presented here indicate that the relative sink strength of a fruit depends upon its age and the number of seeds it contains. A fruit with many seeds retards vegetative growth and leads to the abortion of flower buds, young fruits, and older fruits with fewer seeds. In contrast, a fruit with few seeds permits vegetative growth and flower production and is less dominant over younger fruits. Because embryos and endosperms are known to produce high levels of auxins and other phytohormones, it is possible that hormone production parallels seed number and that the effects of seed number on sink strength are mediated by hormones (Bollard, 1970; Brenner, 1987; Bangerth, 1989).

In conclusion, plants regulate their resource commitment to fruit and seed production under varous environmental conditions via plasticity in seed mass and in the several traits that affect seed number (i.e. flower number, ovules per flower, and fruit and seed abortion). Both the magnitude and the direction of each of these traits is known to vary, at least in some species, in response to environmental conditions. Moreover, in order for plants to regulate maternal investment coherently, suites of traits often change in concert. For example, in the N addition experiment, both seed number per fruit and mean seed mass decreased as total fruit and seed production increased on a plant. However, the directions in which the traits change with respect to one another also vary across environments. For example, in the alternate pollen-load treatment, seed number per fruit and mean seed mass were negatively correlated (in contrast to the N addition experiment).

Recently, a few brave souls have followed the lead of Bradshaw (1965) and have begun to examine the evolution of phenotypic (including reproductive) plasticity in plants (Schlichting, 1986, 1989; Sultan, 1987; Via, 1988). They view the amount of plasticity in a trait (e.g. mean seed mass) and the structure of correlations among traits (e.g. seed number and mean seed mass) across various environments as traits in and of themselves. These plasticity traits, in turn, can potentially respond to evolutionary pressures. As Schlichting (1986) noted, few areas in the plant sciences lie at the intersection of more disciplines (ecology, physiology, developmental morphology, genetics and evolution). To this I will add that few areas are potentially more rewarding to both the pure and the applied plant sciences.

Acknowledgements

I am indebted to L.E. Davis, B. Devlin, J.B. Horton, T.-C. Lau, L. Small, C.D. Schlichting and J.A. Winsor, who did most of the work and provided all of the thought that went into the zucchini experiments; and to T.-C. Lau, T.R. Richardson, O.J. Rocha, and C.D. Schlichting for comments on an earlier draft of this paper. This paper has also benefited from the research supported by the National Science Foundation grants BSR-8314612, BSR-8314612 A01, and BSR-8819184, and by a Hatch Grant from the Pennsylvania State University Agricultural Experiment Station (Project 2683).

References

Adams, M.W. (1967). Basis of yield component compensation in crop plants with special reference to the field bean, *Phaseolus vulgaris*. *Crop Science* **7**, 505–10.

Bangerth, F. (1989). Dominance among fruits/sinks and the search for a correlative signal. *Physiologia Plantarum* **76**, 608–14.

Bawa, K.S. & Webb, C.J. (1984). Flower, fruit, and seed abortion in tropical forest trees: Implications for the evolution of paternal and maternal reproductive patterns. *American Journal of Botany* **71**, 736–51.

Bertin, R.I. (1982). Floral biology, hummingbird pollination and fruit production of trumpet creeper (*Campsis radicans*, Bignoniaceae). *American Journal of Botany* **69**, 122–34.

Bertin, R.I. (1988). Paternity in plants. In *Plant Reproductive Ecology: Patterns and Strategies* (ed. J. Lovett Doust & L. Lovett Doust), pp. 30–50. New York: Oxford University Press.

Black, J.N. (1956). The influence of seed size and depth of sowing on pre-emergence and early vegetative growth of subterranean clover (*Trifolium subterraneum* L.). *Australian Journal of Agricultural Research* **7**, 98–109.

Bollard, E.G. (1970). The physiology and nutrition of developing fruits. In *The Biochemistry of Fruits and their Products* (ed. A.C. Hume), pp. 387–421. New York: Academic Press.

Bookman, S.S. (1984). Evidence for selective fruit production in *Asclepias*. *Evolution* **38**, 72–86.

Bradshaw, A.D. (1965). Evolutionary significance of phenotypic plasticity in plants. *Advances in Genetics* **13**, 115–55.

Brenner, M.L. (1987). The role of hormones in photosynthate partitioning and seed filling. In *Plant Hormones and their Role in Plant Growth and Development* (ed. P.J. Davies), pp. 474–93. Minneapolis: M.N. Shoff.

Burley, N. (1980). Clutch overlap and clutch size: alternative and complimentary reproductive tactics. *American Naturalist* **115**, 223–46.

Charlesworth, D. (1988). Evidence for pollen competition in plants and its relationship to progeny fitness: a comment. *American Naturalist* **132**, 298–302.

Cody, M.L. (1966). A general theory of clutch size. *Evolution* **20**, 174–84.

Davis, L.E. (1986). *Regulation of offspring quality in Cucurbita pepo (Cucurbitaceae)*. Master's thesis. Pennsylvania State University, University Park.

Devlin, B. (1989). Components of seed and pollen yield of *Lobelia cardinalis*: variation and correlations. *American Journal of Botany* **76**, 204–14.

Galen, C., Plowright, R.C. & Thomson, J.D. (1985). Floral biology and regulation of seed set and seed size in the lily *Clintonia borealis* (Ait.) Raf. *American Journal of Botany* **72**, 1544–52.

Giesel, J.T. (1976). Reproductive strategies as adaptations to life in temporally heterogeneous environments. *Annual Review of Ecology and Systematics* **7**, 57–80.

Goldschmidt, E.E. & Golomb, A. (1982). The carbohydrate balance of alternate-bearing citrus trees and the significance of reserves for flowering and fruiting. *Journal of the American Society for Horticultural Science* **107**, 206–8.

Gross, K.L. (1984). Effects of seed size and growth form on seedling establishment of six monocarpic perennial plants. *Journal of Ecology* **72**, 369–87.

Harper, J.L. & White, J. (1974). The demography of plants. *Annual Review of Ecology and Systematics* **5**, 419–63.

Howe, H.F. (1976). Egg size, hatching asynchrony, sex, and brood reduction in the common grackle. *Ecology* **57**, 1195–207.

Lack, D. (1954). *The Natural Regulation of Animal Numbers*. Oxford: Clarendon Press.

Lack, D. (1966). *Population Studies of Birds*. Oxford: Clarendon Press.

Lee, T.D. (1988). Patterns of fruit and seed production. In *Plant Reproductive Ecology: Patterns and Strategies* (ed. J. Lovett Doust & L. Lovett Doust), pp. 179–96. New York: Oxford University Press.

Lee, T.D. & Bazzaz, F.A. (1982a). Regulation of fruit and seed production in an annual legume, *Cassia fasciculata*. *Ecology* **63**, 1363–73.

Lee, T.D. & Bazzaz, F.A. (1982b). Regulation of the fruit maturation pattern in an annual legume, *Cassia fasciculata*. *Ecology* **63**, 1374–88.

Levin, D.A. (1984). Inbreeding depression and proximity-dependent crossing success in *Phlox drummondii*. *Evolution* **38**, 116–27.

Lloyd, D.G. (1980). Sexual strategies in plants. I. An hypothesis of serial adjustment of maternal investment during one reproductive session. *New Phytologist* **86**, 69–80.

Lovett Doust, J. & Eaton, G.W. (1982). Demographic aspects of flower and fruit production in bean plants, *Phaseolus vulgaris* L. *American Journal of Botany* **69**, 1156–64.

Low, B.S. (1978). Environmental uncertainty and the parental strategies of marsupials and placentals. *American Naturalist* **112**, 197–213.

Marshall, D.L., Levin, D.A. & Fowler, N.L. (1985). Plasticity in yield components in response to fruit predation and data of fruit initiation in three species of *Sesbania* (Leguminosae). *Journal of Ecology* **73**, 71–80.

Marshall, D.L., Levin, D.A. & Fowler, N.L. (1986). Plasticity of yield components in response to stress in *Sesbania macrocarpa* and *Sesbania vesicaria* (Leguminosae). *American Naturalist* **127**, 508–21.

Mazer, S.J. (1987). The quantitative genetics of life history and fitness components in *Raphanus raphanistrum* L. (Brassicaceae): ecological and evolutionary consequences of seed-weight variation. *American Naturalist* **130**, 891–914.

Mitchell, R. (1977). Bruchid beetles and seed packaging by palo verde. *Ecology* **58**, 644–51.

Nakamura, R.R. (1988). Seed abortion and seed size variation within fruits of *Phaseolus vulgaris*: Pollen donor and resource limitation effects. *American Journal of Botany* **75**, 1003–10.

Primack, R.B. (1978). Regulation of seed yield in *Plantago*. *Journal of Ecology* **66**, 835–47.

Rocha, O.J. & Stephenson, A.G. (1990). Effect of ovule position on seed production, seed weight and progeny performance in *Phaseolus coccineus* L. (Leguminosae). *American Journal of Botany* **77**, 1320–9.

SAS Institute (1982). *SAS User's Guide: Statistics*. Cary, North Carolina: SAS Institute.

Schaal, B.A. (1980). Measurement of gene flow in *Lupinus texensis*. *Nature* **284**, 450–1.

Schaffer, A.A., Goldschmidt, E.E., Goren, R. & Galili, D. (1985). Fruit set and carbohydrate status in alternate and nonalternate bearing citrus cultivars. *Journal of the American Society for Horticultural Science* **110**, 575–8.

Schlessman, M.A. (1988). Gender diphasy ('sex choice'). In *Plant Reproductive Ecology: Patterns and Strategies* (ed. J. Lovett Doust & L. Lovett Doust), pp. 139–70. New York: Oxford University Press.

Schlichting, C.D. (1986). The evolution of phenotypic plasticity in plants. *Annual Review of Ecology and Systematics* **17**, 667–93.

Schlichting, C.D. (1989). Phenotypic integration and environmental change. *BioScience* **39**, 460–4.

Schlichting, C.D. & Levin, D.A. (1984). Phenotypic plasticity in annual *Phlox*: tests for some hypotheses. *American Journal of Botany* **71**, 252–60.

Schlichting, C.D., Stephenson, A.G., Small, L.E. & Winsor, J.A. (1990). Pollen loads and progeny vigor in *Cucurbita pepo*: the next generation. *Evolution* **44**, 1358–72.

Small, L.E. (1988). *The effects of pollen tube competition on the second generation in Cucurbita pepo (Cucurbitaceae)*. Master's thesis, Pennsylvania State University, University Park.

Snow, A.A. & Mazer, S.J. (1988). Gametophytic selection in *Raphanus raphanistrum*: a test for heritable variation in pollen competitive ability. *Evolution* **42**, 1065–75.

Solomon, B.P. (1986). Sexual allocation and andromonopecy: Resource investment in male and hermaphroditic flowers of *Solanum carolinense* (Solanaceae). *American Journal of Botany* **73**, 1215–21.

Sparks, D. & Madden, G. (1985). Pistillate flower and fruit abortion in Pecan as a function of cultivar, time, and pollination. *Journal of the American Society for Horticultural Science* **110**, 219–23.

Stanton, M.L. (1984). Seed variation in wild radish: Effect of seed size on components of seedling and adult fitness. *Ecology* **65**, 1105–12.

Stanton, M.L., Bereczky, J.K. & Hasbrouck, H.D. (1987). Pollination thoroughness and maternal yield regulation in wild radish, *Raphanus raphanistrum* (Brassicaceae). *Oecologia* **74**, 68–76.

Stearns, S.C. (1976). Life-history tactics: a review of the ideas. *Quarterly Review of Biology* **51**, 3–47.

Stephenson, A.G. (1981). Flower and fruit abortion: proximate causes and ultimate functions. *Annual Review of Ecology and Systematics* **12**, 253–79.

Stephenson, A.G. (1984). The regulation of maternal investment in an indeterminate flowering plant (*Lotus corniculatus*). *Ecology* **65**, 113–21.

Stephenson, A.G., Devlin, B. & Horton, J.B. (1988a). The effects of seed number and prior fruit dominance on the pattern of fruit production in *Cucurbita pepo* (zucchini squash). *Annals of Botany* **62**, 653–61.

Stephenson, A.G. & Winsor, J.A. (1986). *Lotus corniculatus* regulates offspring quality through selective fruit abortion. *Evolution* **40**, 453–8.

Stephenson, A.G., Winsor, J.A., Schlichting, C.D. & Davis, L.E. (1988b). Pollen competition, nonrandom fertilization, and progeny fitness: A reply to Charlesworth. *American Naturalist* **132**, 303–8.

Sultan, S.E. (1987). Evolutionary implications of phenotypic plasticity in plants. *Evolutionary Biology* **21**, 127–78.

Sutherland, S. & Delph, L.F. (1984). On the importance of male fitness in plants: Patterns of fruit-set. *Ecology* **65**, 1093–104.

Thomson, J.D. (1989). Deployment of ovules and pollen among flowers within inflorescences. *Evolutionary Trends in Plants* **3**, 65–8.

Vanderplank, F.L. (1958). Studies on the coconut pest, *Pseudotheraptus wayi* Brown (Coreidae), in Zanzibar. I. A method of assessing the damage caused by the insect. *Bulletin of Entomological Research* **49**, 559–84.

Vanderplank, F.L. (1960). Studies on the coconut pest, *Pseudotheraptus wayi* Brown (Coreidae), in Zanzibar. II. Some data on the yields of coconuts in relation to damage caused by the insect. *Bulletin of Entomological Research* **50**, 135–49.

van Steveninck, R.F.M. (1957). Factors affecting the abscission of reproductive organs in yellow lupins (*Lupinus luteus* L.). I. The effect of different patterns of flower removal. *Journal of Experimental Botany* **8**, 373–81.

Vasek, F.C., Weng, V., Beaver, R.J. & Huszar, C.K. (1987). Effects of mineral nutrition on components of reproduction in *Clarkia unguiculata*. *Aliso* **11**, 599–618.

Via, S. (1988). Genetic constraints on the evolution of phenotypic plasticity. In *Genetic Constraints on Adaptive Evolution* (ed. V. Loeschke), pp. 47–71. Berlin: Springer-Verlag.

Watson, M.A. & Casper, B.B. (1984). Morphogenetic constraints on patterns of carbon distribution in plants. *Annual Review of Ecology and Systematics* **15**, 233–58.

Wiens, D. (1984). Ovule survivorship, brood size, life history, breeding system, and reproductive success in plants. *Oecologia* **64**, 47–53.

Willson, M.F. & Price, P.W. (1980). Resource limitation of fruit and seed production in some *Asclepias* species. *Canadian Journal of Botany* **58**, 2229–33.

Winsor, J.A., Davis, L.E. & Stephenson, A.G. (1987). The relationship between pollen load and fruit maturation and the effect of pollen load on offspring vigor in *Curcurbita pepo*. *American Naturalist* **129**, 643–56.

C. MARSHALL and M. A. WATSON

Ecological and physiological aspects of reproductive allocation

Introduction

The pattern of resource allocation in plants and its regulation is of great practical importance to agriculturalists and horticulturalists, and is of conceptual interest to ecologists and evolutionary biologists. This chapter discusses ecological and physiological aspects of reproductive allocation and includes an approach to allocation based on plant developmental morphology; the improvement of agricultural yield in terms of increased allocation to reproductive sinks is also evaluated.

It is generally considered that nutritional and energetic resources are allocated in a strategic manner that maximizes fitness (Harper, 1977; Fitter, 1986; Bazzaz *et al.*, 1987). Resources are allocated to three activities: growth, reproduction and defence. The inherent pattern of allocation is considered to reflect the effect of past selection pressures in optimizing both fecundity and survival. Thus genetic differences in the pattern of resource allocation tend to be broadly associated with different ecological strategies. For example, Harper (1977) has argued that in a crowded community where there will be intense competition for resources a successful individual needs vigorous shoot and root growth to maximize resource acquisition; in this case fecundity will be associated with, and depend upon, a major allocation of resources to vegetative growth. In contrast in more open environments with relatively little competition from neighbouring individuals allocation to vegetative growth becomes less important and so a large proportion of the total resource may be invested in reproductive growth and development, thereby maximizing the chance of leaving descendants. In addition, there are situations where predators will be a major selective element and correspondingly resources become diverted from growth to defence via the production of secondary metabolites or structures such as thorns and prickles. In these cases fecundity will be reliant upon the degree of investment in defence.

Overall, the investment of resources to one set of functions seems likely to lead to a decrease in allocation to other types of organs or functions, implying, for example, that reproductive growth and development must occur at the expense of vegetative growth and development. Indeed there is evidence that a high allocation to reproduction in iteroparous perennials is often correlated with reduced vegetative performance, decreased potential for future reproduction, and decreased longevity (Eis, Garman & Ebel, 1965; Harper, 1977; Sohn & Policansky, 1977; Law, 1979). The cost of reproduction must therefore be evaluated in terms of future survival, i.e. if seed is produced too early or too profusely the vegetative cost may be a reduction in vigour or competitive ability and correspondingly a higher risk of plant mortality. In monocarpic plants, according to this view, the cost of reproduction is death, because of the 'self-destruction' of annual plants due to the extensive allocation of nutrients to fruit and seed development (Sinclair & de Wit, 1975).

Measurement of resource allocation

Underlying problems

There are considerable difficulties in quantifying the allocation of resources to different functional areas. With respect to reproductive allocation, differences in life history such as the precocity of reproductive onset or the number of reproductive events (semelparity versus iteroparity) confound comparisons of species. There is also the fundamental problem that the reproductive structures of many plants make a major contribution, by photosynthesis, to their growth and development, i.e. they are not wholly dependent on the allocation of carbon from the vegetative structure of the plant (Bazzaz, Carlson & Harper, 1979; Watson & Casper, 1984; Bazzaz & Reekie, 1985). For example, Bazzaz & Carlson (1979) estimated that around 50% of the carbon utilized in the production of reproductive structures in *Ambrosia trifida* was contributed by their own photosynthesis. In temperate cereals it is well established that ear photosynthesis makes a significant contribution to total canopy photosynthesis and correspondingly to seed growth (Evans & Wardlaw, 1976). For example, Biscoe *et al.* (1975) recorded that photosynthesis of the awned ear of spring barley contributed 18% of total canopy photosynthesis from anthesis onwards; this was equivalent to 13% of the final grain mass. Similarly, in grain legumes, pod photosynthesis makes a substantial contribution to seed growth; a major component of this activity is the re-assimilation of carbon released within the pod by seed respiration (Flinn & Pate, 1970; Pate, 1984). These examples emphasize that a significant proportion of the carbon requirements of reproductive structures is not necessarily gained at the expense of vegetative development. In addition, as the major mineral nutrients (N, P and K) are recycled during development from old to young tissues via the phloem, their distribution to reproductive tissues is not wholly at the expense of vegetative growth (see p. 190). Nevertheless all the water consumed by reproductive parts can be costed against its potential utilization in vegetative growth. The interpretation of patterns of allocation of resources is thus far from straightforward.

Allocation currency

Similarly, there are considerable difficulties in determining the most appropriate currency for evaluating allocation, and further, how it should be applied. Ideally allocation should be based on the limiting resource, but this is not readily identified and has the potential drawback that it is likely to vary with time (stage of development) and with the environment. The simplest and most widely used currency is biomass (Fig. 1*b*,*c*) and this is often equated with carbon (Fitter, 1986). The proportion of total biomass that is associated with reproductive structures is easily measured, but it is not a precise measure of carbon allocation as it does not distinguish between energy-poor and energy-rich compounds or between locally produced and transported carbon. In view of this it has been proposed that carbon itself is a far more appropriate currency than biomass (Watson & Casper, 1984). Bazzaz & Reekie (1985) have suggested that carbon, as distinct from biomass, could act as the currency integrating the allocation of other resources. For example, there is likely to be a close relationship between the overall supply of carbon to the root system and the uptake of mineral nutrients from the soil. Alternatively, the energy content of reproductive structures has been used as currency (Fig. 1*a*); as might be expected, this frequently gives a broadly similar pattern of allocation to that of biomass (Harper & Ogden, 1970; Hickman & Pitelka, 1975; Abrahamson & Caswell, 1982). Mineral nutrients, e.g. nitrogen and phosphorus, have also been used as currency for investigations of resource allocation (Van Andel & Vera, 1977; Lovett Doust, 1980; Abrahamson & Caswell, 1982; Fitter & Setters, 1988) and often show quite different allocations to reproductive structures compared with biomass (Fig. 1*c*). In addition, each major nutrient appears to make a different proportional contribution to reproductive structures, and so a currency based on mineral allocation is not easy to interpret. However, despite all these problems, the relative order of reproductive allocation between different species and populations appears to be broadly similar irrespective of the currency that is used for the comparison.

Thus, despite its obvious limitations, biomass distribution provides a

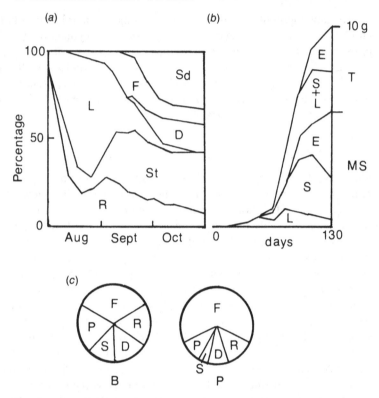

Fig. 1. Allocation of energy, biomass and phosphorus among various plant parts. (*a*) Percentage allocation of calories through the life cycle of *Senecio vulgaris* (after Harper & Ogden, 1970). Sd, seeds; F, flowers and receptacles; L, live leaves; D, dead leaves; St, stem and hypocotyl; R, roots. (*b*) Distribution of biomass through the life cycle of a biculm selection of wheat (after Marshall & Boyd, 1985). MS, main shoot; T, tiller; L, leaves; S, stem; E, ear. (*c*) Distribution of biomass (B) and phosphorus (P) after 79 weeks in *Smyrnium olusatrum* (after Lovett Doust, 1980). F, flowers and fruits; P, peduncles; S, stem; D, dead leaves; R, tuberous and fibrous roots.

general approach for allocation studies; but it is important that equivalent comparisons are made, e.g. in relation to the inclusion of flower parts and supporting structures as well as seeds and fruits. In this respect Thompson & Stewart (1981) proposed that the term *reproductive effort* (RE) should be used to cover investment in all reproductive structures. In practice, however, especially in the context of the yield of crop plants, only the biomass of the harvested seed or fruit is normally considered, and the proportion of the total above-ground biomass that is represented by the seed or fruit yield is known as the harvest index (HI). Thus, as root biomass is usually ignored, HI values give an overestimate of true reproductive allocation. On the other hand, the final biomass of reproductive structures does not take into account the respiratory costs associated with their growth and maintenance and so is a far from absolute measure of their actual consumption of carbon.

Physiological aspects of allocation

In physiological terms, the distribution of biomass between the various parts of a plant represents the balance between the total partitioning of assimilate from source leaves to sinks throughout development and its consumption by respiration in sink tissues. Differences in allocation between sinks for the most part reflect differences in their relative competitive ability, i.e. differences in sink strength: the product of sink size and sink activity (Warren Wilson, 1972; Ho, this volume). The source–sink system is physiologically integrated and appears to be regulated predominantly by the sink, but the critical metabolic and physiological properties that underlie sink strength and thereby influence the flow of assimilate and other materials to sink tissues remain uncertain. Nevertheless there is evidence that hormonal influences are particularly important in regulating the sink activity of developing fruits and seeds (Brenner, 1988; Bangerth, 1989).

In most cases there is competition for assimilate between newly produced reproductive structures and vegetative growth, but as fruit development progresses reproductive growth becomes increasingly dominant and consequently vegetative growth declines. Thus, for example, in annual dicotyledons, fruit and seed production is normally associated with a decline in both root growth and bud outgrowth as well as the onset of leaf senescence. Correspondingly, the removal of growing fruit or the dissection of developing seeds from fruits results in the resumption of vegetative growth and the decline or even reversal of leaf senescence (Wareing & Phillips, 1981; Tamas *et al.*, 1979*a*, 1981; Nooden & Guiamet, 1989). However, both the decline in axillary bud growth and the onset of senescence in fruiting plants appear to be regulated by hormonal substances released by developing seeds (with IAA probably playing the major role) (Tamas *et al.*, 1979*a*, 1981). Similarly, in temperate grasses and cereals, the characteristic decline in tillering that occurs during the early phase of stem elongation is also regulated hormonally, for removal of the unemerged developing inflorescence or application of an IAA transport inhibitor such as TIBA readily restores tillering capacity

(Jewiss, 1972; Clifford, 1977; Harrison & Kaufman, 1980). These examples of correlative inhibition emphasize that the acquisition of resources by newly developing reproductive sinks is closely related to underlying hormonal changes that tend to suppress vegetative growth and development. In this way the potential of vegetative sinks is reduced and so reproductive sinks become predominant in competitive inter-relations.

As well as the initial competition between vegetative and reproductive growth, in many cases there is also competition between individual fruits that leads to differences in fruit size and in cases of severe competition to the abortion of developing fruits (see chapters by Goldwin, Ho, and Stephenson, this volume). In general it appears that the fruits that set first (i.e. the oldest) exert dominance over those that set later on; thus the latter tend to be smaller and are more likely to abort (Tamas *et al.*, 1979*b*; Stephenson, 1981). For several species there appears to be a fairly close correlation between the physiological dominance (sink strength) of individual fruits and seed number, and this seems to be related to the underlying hormonal status (Stephenson, Devlin & Horton, 1988; Stephenson, this volume; Bangerth, 1989).

In general, the yield of fruit and seed is set by the resource supply rather than by the production of potential reproductive sinks: for it is a characteristic feature of plant reproductive biology that the number of flowers produced greatly exceeds the number that set fruit, and this in turn exceeds the number of fruits that reach maturity (see Goldwin, this volume). However, such losses in reproductive potential are not always attributable to resource competition. Many outbreeding species display a high degree of ovule and seed abortion (Wiens, 1984; Wiens *et al.*, 1987; Marshall & Ludlam, 1989). For example, in the inflorescence of *Lolium perenne*, Marshall & Ludlam (1989) observed that around 50% of the florets were unproductive, as they aborted ovaries and very small developing seeds. There was a similar degree of loss from all florets, i.e. abortion was not associated with age or positional hierachies within spikelets, and treatments designed to alleviate resource competition within the inflorescence had no effect on the degree or pattern of abortion. Thus if competition for resources is not a major influence regulating abortion, then it is possible that genetic defects in ovules that are associated with outbreeding give rise to the very high degree of seed abortion in herbage grasses and other outbreeding species (Wiens *et al.*, 1987). Thus the breeding system itself, rather than the supply of resources, may limit reproductive performance in such plants.

Resource allocation and plant life history

The allocation of biomass to sexual reproduction, often termed reproductive allocation (RA) by ecologists, in plants with different life history appears to follow a general trend from a high RA in annuals and monocarpic perennials (of the order of 25–50%) to a lower RA (of the order of 5–25%) in iteroparous perennials (Harper, 1977; Abrahamson, 1979; Fitter, 1986). An analysis of RA in congeneric and conspecific annuals and perennials by Bazzaz *et al.* (1987) showed that the latter had consistently lower values than the former group. Similar selection pressures acting on individuals in similar environments appear to result in common patterns of resource allocation. This closely relates to the theory of *r*- and *K*-selected plants with selection acting on resource allocation (Fitter, 1986), and with Grime's (1979) triangular model of plant strategy in relation to habitat stability and environmental favourability. Thus annuals are generally associated with relatively unstable habitats and have a high reproductive allocation giving a high potential rate of population increase (*r*-selection). Conversely, species associated with more predictable habitats normally face intense competition and have a lower reproductive allocation (*K*-selection), thereby directing resources to prolonged survival. These general differences in life history strategy are also illustrated by observations on closely related species and populations from variable and stable environments. For example, species of *Solidago* from contrasting habitats show a marked decline in reproductive effort (from around 0.45 to 0.10) with transition from less mature to more mature habitats, and this change was inversely related to plant size (Abrahamson & Gadgil, 1973). Law, Bradshaw & Putwain (1977) showed similar differences between populations of *Poa annua* originating from contrasting habitats. These results give support to the generalization that the inherent pattern of resource allocation between vegetative and reproductive structures is closely related to habitat, although there are some notable exceptions (Watson & Casper, 1984). However, the general relationship between allocation strategy and the presumed selective forces that predominate in different habitats is wholly correlative and needs to be tested by experimental investigation (Primack & Antonovics, 1982; Davy & Smith, 1988). Reciprocal transplantation of closely related populations between alien sites or into common gardens perhaps offer experimental approaches that allow a more realistic evaluation of the effect of selection pressures on resource allocation and plant survival. However, few definitive studies have as yet been made in this area.

Plasticity of resource allocation

There are further difficulties in making an absolute assessment of the allocation of resources to reproduction in different species, as the pattern of allocation shows marked plasticity. In particular, when resources become limiting, reproduction may be either suppressed or promoted (Harper & Ogden, 1970; Pitelka, Stanton & Peckenham, 1980; Grime, Crick & Rincon, 1986). In ephemeral annuals at high plant densities, where growth rate and plant size are likely to be restricted by shading and the supply of water and mineral nutrients, many individuals may not even survive to maturity. However, those that do commonly display an earlier onset of reproductive development and tend to sustain a relatively high RA; nevertheless this is substantially lower than that of spaced individuals (Hickman, 1975; Harper, 1977). In *Plantago coronopus*, which displays an annual strategy under experimental conditions, Waite & Hutchings (1982) showed that RA significantly decreased with increasing density (from 47 to 31%) and that this this was correlated with a significant reduction in shoot biomass. Similarly, in temperate cereals the harvest index per unit crop area declines as density increases (Donald & Hamblin, 1976) and this reflects the different responses of total biomass and grain yield to increase in plant density. Whereas the former increases with density to reach a plateau, grain yield reaches a maximum and then declines with further increase in density. It is generally assumed that this decline in grain production at very high density is related to the earlier competition for light experienced at high densities; this in turn results in a higher proportion of infertile tillers and fewer and smaller grains per ear. Shading crop canopies similarly reduces the harvest index (Fischer & Wilson, 1975).

Perennial species of stable habitats show a similar reduction in RA at high density but they may also defer reproduction and thereby maximize survival by conserving the potential for future reproduction (Grime *et al.*, 1986). In species such as *Tussilago farfara*, in which increase in population size may result from both vegetative growth and sexual reproduction, the balance between these activities changes markedly with increasing density. Sexual reproduction appears to be less sensitive than clonal expansion to reduction in resource supply (Ogden, 1974).

A morphological approach to resource allocation

From the preceding discussion, it is evident that studies of the determinants of patterns of resource allocation in plants are typically based on an assumption that either energy (measured as carbon or dry mass) or mineral nutrients are the relevant resources that are limiting to plant growth. This assumption reflects, at least in part, the origins of life-history theory in the animal literature, where it is suggested that competition exists among the processes of reproduction, growth and maintenance for limited energy resources (Cody, 1966). Thus, as outlined in the Introduction, resources used to support one type of activity are unavailable to support another, resulting in trade-offs in the allocation of critical limiting resources to organs that differ in life-history function.

However, Harper (1977) suggests that 'the green plant may indeed be a pathological overproducer of carbohydrates', implying that energy, and by extension photosynthetic assimilate, may not be a critical limiting resource in plants (Watson & Casper, 1984). This may result from the fact that plants differ from animals in some important ways. Plant growth is modular, resulting in the ordered production of vegetative and sexual shoot systems or modules (White, 1979, 1984). These components may be photosynthetic, and thus partly or entirely capable of paying their own carbon costs (see above for discussion), so that they are not simply resource drains on the system. Plants are effectively able to increase their number of 'mouths'. However, the component modules are not equivalent in these abilities and, depending on the type of module that is produced, the efficiency of that module as a mouth may vary (Watson, 1984).

Thus, while patterns of resource allocation in plants reflect the types, and frequencies of types, of modules that are formed (Maillette, 1982; Tuomi, Niemela & Mannila, 1982), these patterns do not necessarily result from an active allocation of limiting resources. A trade-off among modules with differing life-history functions should neither be expected nor implied (see also Tuomi, Makala and Haukioja, 1983). The plant growth form that ultimately results, however, should reflect evolutionarily derived controls on programmes of plant development. These controls may or may not be directly responsive to the immediate resource environment.

An example of a system in which developmental controls on flowering do not appear to be directly responsive to the resource environment can be found in some strictly biennial varieties of apple. In these varieties, new floral buds are not initiated on current fruit-bearing branches (Chan & Cain, 1967). Typically, this would be interpreted as resulting from the high energetic costs of fruit production: a plant that fruits one year lacks the resources to fruit the next (see Goldwin, this volume). However, at least in some varieties, the failure to form new flower buds is due to the presence on the branch of developing seeds rather than developing fruit. Individual shoots of a parthenocarpic variety of apple that normally bear fruit annually can be made to bear biennially solely by pollinating the flowers. Pollination results in seed formation. No flower buds are initiated on shoots that are occupied by fruits that develop from pollinated flowers. Growth substances produced by the developing seeds, rather than local depletion of resources by the developing fruits, appear to be responsible for the suppression of floral bud formation (Chan & Cain, 1967). Although this developmental programme may have evolved to prevent long-term depletion of resources caused by repeated fruit formation in the natural environment, the programme does not appear to be triggered by resource depletion itself.

Thus, if carbon (and perhaps mineral nutrients) are not limiting, the question is raised, is there an alternative currency to carbon that may be more appropriate for assessing the role of environmental pressures on the evolution of plant form and the expression of plant productivity? Below, we suggest that the functional meristem pool may be such an appropriate limiting resource, at least in some cases.

Viable meristems as a potentially significant limited currency

A number of recent studies have demonstrated that a plant's architecture and developmental pattern can place significant constraints on its ability to respond to environmental change. In the water hyacinth, *Eichhornia crassipes* Solms., for example, onset of inflorescence production during the exponential phase of ramet population increase leads to the decreased production of ramets (Fig. 2) (Watson, 1984). This is precisely the pattern of trade-off predicted by traditional life-history theory. However, the observed trade-off between sexual and vegetative growth is more easily explained by examination of the developmental programme of the plant than by examination of its carbon or nutrient economy. In water hyacinth, there are two separate meristem populations, the apical and the lateral, that differ in their potential developmental fates (Richards, 1980; Watson & Cook, 1982; Watson, Carrier & Cook, 1982). Shoot apical meristems may (i) differentiate leaves and associated lateral meristems, resulting in vertical growth of the parental axis; (ii) remain quiescent, in which case the parent axis ceases to increase in size; or (iii) differentiate inflorescences, with the result that the shoot axis becomes determinate. Lateral meristems also have three potential developmental fates. They may differentiate new ramets, leading to an increase in size of the ramet population; they may remain quiescent and ultimately abort, in which case there is no increase in the ramet population; or, if the shoot apical meristem flowers, they may form a continuation shoot (Richards, 1980).

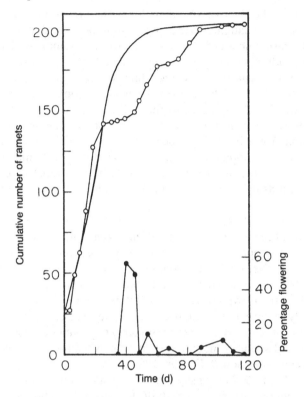

Fig. 2. Patterns of ramet production and flowering with time in *Eichhornia crassipes* (after Watson, 1984). Open circles, cumulative number of ramets; filled circles, percentage of flowering ramets. Smooth curve is fitted logistic function for ramet population.

The latter continues vertical extension of the flowering ramet, so that its growth is functionally indeterminate. The outcome of this programme of meristem determination is that the growth of the parental axis is maintained at the 'expense' of daughter ramet production and, hence, of clonal spread.

This pattern of development, which is quite independent of the plant's carbon economy, explains the trade-off between ramet production and flowering observed in experimental populations (Watson, 1984). During early expansion of water hyacinth populations, all newly formed lateral meristems develop into daughter ramets, resulting in an exponential increase in the size of the ramet population. In exponentially growing populations that begin to flower, however, each inflorescence produced by a ramet causes the removal of a lateral meristem from the pool available to that ramet for daughter ramet production, because that meristem forms a continuation shoot. Given constant rates of ramet production, inflorescence production must automatically result in a decreased rate of ramet production, regardless of the state of energetic or nutrient resources in the plant (Watson, 1984; Watson & Cook, 1987; Geber, Watson & Furnish, 1992). Thus, in these plants, patterns of resource allocation appear to reflect the outcome of competition among life-history functions for a limited number of meristems. Commitment of a meristem to one developmental pathway precludes its commitment to another. The choice has important life-history consequences to the plant that are manifest as differences in growth form and patterns of resource allocation.

Although Watson (1984) demonstrated that carbon need not be limiting in order to observe a trade-off between sexual and vegetative function, this did leave open the possibility that production of one type of module versus another might have long-term consequences to the carbon economy of the plant. Even though sexual branches may be able to support much of their own growth and maintenance, their formation still may ultimately reduce the amount of assimilate available for storage, because successful sexual structures are unlikely to produce excess assimilate. Thus, sexual shoot or module formation may well influence carbon balance indirectly (Watson & Casper, 1984).

A study by Sohn & Policansky (1977) on mayapple, *Podophyllum peltatum*, a common rhizomatous herb of the northeast deciduous forest of the United States, reflected the commonly held belief that sexual branches are more expensive to produce than vegetative ones. Sohn & Policansky showed, by way of a matrix projection model based on demographic data, that repeated sexual shoot formation would ultimately lead to clone extinction. However, it remained unclear whether, if their projection was correct, it could be explained by the differential production cost of the two shoot types (Watson & Casper, 1984; Benner & Watson, 1989; Watson, 1990). Sexual and vegetative shoots of mayapple differ significantly in architecture (Fig. 3). The two-leaved sexual shoot has about twice the leaf area and weighs approximately twice as much as the single-leaved vegetative shoot. Sohn & Policansky (1977) suggested that the size difference alone should result in a higher cost of sexual shoot production. However, it is also possible that both shoot types quickly become self-supporting with respect to carbon, thus diminishing any real differences in the carbon cost of production between them. Furthermore, differences in the future contributive capacity of the two shoot types also may be minimized because of the evolution of the two distinct shoot

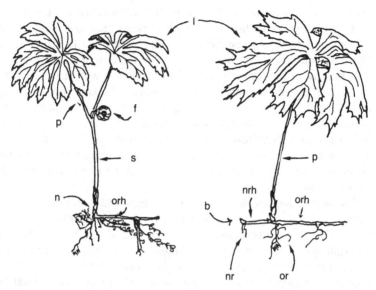

Fig. 3. Ramets of mayapple (*Podophyllum peltatum*) (after Watson, 1990). Sexual ramet on left, vegetative ramet on right; orh, old rhizome; n, node; or, old roots; p, petiole; s, stem; l, leaf lamina; f, flower; nrh, new rhizome; nr, new roots; b, new bud.

architectures. The single-leaved vegetative shoot and the dual-leaved sexual shoot each appear capable of generating enough photosynthetic assimilate to support their respective demographic roles – in the vegetative shoot this includes leaf development and maintenance, and fruit production in the sexual shoot – while still generating sufficient resources to produce perennating organs of equivalent size to ensure continued growth by the clone (Sohn & Policansky, 1977; Watson, 1990).

Why, then, should continued sexual shoot formation lead to clone extinction? Interestingly, the answer may lie in the dynamics of the quiescent meristem population of the perennial rhizome system. These quiescent meristems give rise to new rhizome branches in the event of damage to the terminal portion of the rhizome system, and are of two sorts, apical and lateral. Vegetative shoots, which are actually leaves (Holm, 1899), develop without consuming the apical meristem of the sympodial shoot; in sexual branches the apex is consumed (Fig. 3). Thus, old rhizome segments that gave rise to sexual shoots contain only quiescent lateral buds, whereas old rhizome segments that gave rise to vegetative shoots bear both lateral and apical buds. Preliminary observations suggest that when damage to forward-growing rhizome segments occurs, bud release takes place at nodes no more than two nodes proximal to the point of damage, and it appears that the quiescent apical meristems left by old vegetative shoots are the first to grow out (M.A. Watson, unpublished). Thus, it seems that the cost of sexual shoot formation may be a reduced capacity of the rhizome system to recover from damage. Rhizome systems containing many old sexual-shoot-bearing segments, or those that bore old sexual shoots at locations just proximal to points of injury, may be slower to respond to damage. If experiments that are in progress demonstrate a difference in the ability of former sexual and vegetative rhizome segments to recover from damage, there may be an indirect but very real cost of sexual shoot formation in *P. peltatum* over the long term.

The importance of developmental pattern and its interaction with the meristem pool is not restricted to clonal plants. Geber (1989, 1990) found, in a study of two co-occurring *Polygonum* species, that the pattern and timing of commitment of lateral meristems to vegetative branching or flower production varied within and between the two species. Between-species comparisons revealed that, although the developmental framework was similar for both species, the programme of meristem determination varied, leading to differences in plant form and productivity between them (Geber, 1989). Comparisons within *Polygonum arenastrum* demonstrated a negative correlation between fecundity and growth that was related to the timing of commitment of lateral meristems to sexual versus vegetative functions, and had a strong genetic component (Geber, 1990). Substantial genetic variation in the expression of developmental programme also has been observed in water hyacinth, where it has been shown to effect the differential competitive abilities of clones (Watson, 1987; Geber *et al.*, 1992).

Phenology as a developmental constraint on allocation

Little work has been done to examine how the phenology of developmental events within a plant, such as the timing of determination of particular buds or meristems, affects how a plant responds to environmental change to give a particular pattern of resource allocation. A simple example can be found in water hyacinth, where inflorescence production has profoundly different effects depending upon whether it occurs during exponential expansion of the ramet population or after the population has reached stationary phase (Watson, 1984). Inflorescence production during the early stages of population growth, as discussed above, results in significantly decreased rates of ramet production, while inflorescence production at later stages of population growth exerts little effect (Watson, 1984; Watson & Cook, 1987; Geber *et al.*, 1992). An appreciation for the importance of the *timing* of inflorescence induction on allocation to vegetative growth, and hence increase in ramet population, could be significant in searching for alternative effective biological control for this noxious weed. A strategy aimed at inducing flowering early in population growth, rather than later on, could effectively curtail the population expansion of water hyacinth. Unfortunately, there is little understanding of the factors that induce flowering in this species (Watson & Brochier, 1988).

In perennial plants that maintain populations of pre-formed buds (i.e. buds that contain pre-determined structures that will grow out at a later date), there is the potential for more complex interactions, between the timing of bud determination and the pattern of environmental variability, which may or may not result in 'adaptive' responses. Plants with this form of developmental phenology are common on the deciduous forest floor of the northeastern United States (Foerste, 1891); but the growth form is also characteristic of a large number of trees and shrubs, many of which are commercially important, such as the parthenocarpic apples discussed above.

In many plants, onset of flowering is strongly correlated with individual plant size. However, detailed studies by Pitelka and co-workers on *Aster acuminatus*, another perennial herb of the northeast deciduous forest (summarized in Watson, 1990), indicate that not all plants of a given size necessarily flower, despite the fact that the probability of flowering increases significantly with plant size and level of irradiance. In a structured experiment, they showed that it was not the total amount of light received *per se* that was significant for the flowering response, but rather the growth stage at which plants were exposed to a period of high irradiance. Exposure early in development greatly promoted overall growth but exposure later on resulted in plants that flowered more. Pitelka, Ashmun & Brown (1985) suggested that the low correlation between plant size and flowering effort was related to the developmental phenology of the plant, in which vegetative growth is completed before flowering begins. They suggested that separation in the timing of development of modules differing in life-history function allows seasonal patterns of environmental variation to exert relatively independent effects on them.

As the time between bud determination and outgrowth grows longer, the early determination of buds becomes an increasing constraint on the abilities of plants to respond to environmental change. In mayapple, the decision regarding whether the next year's shoot will be sexual or vegetative is made during the previous growth season. Initial studies suggested that the decision was made late in the summer preceding the year of outgrowth, presumably allowing the developing apex to assess the resource state of the rhizome system at the end of the growth season (Sohn & Policansky, 1977). Recent studies, however, suggest that the decision to make a vegetative or sexual shoot is made much earlier, perhaps concommitantly with the outgrowth of the current year's shoot (Watson, 1990).

Such a schedule of bud determination would have profound consequences for the plant, because it suggests that the critical developmental decision to make a sexual or vegetative shoot is made before information is complete about the resource status of the terminal end of the rhizome system and in the absence of information about the characteristics of the spatial environment into which the new shoot will emerge in the following year. This pattern of developmental phenology would seem to place a severe constraint on the plant's ability to respond to year-to-year or microenvironmental sources of variability and suggests firstly that the rhizome system must be highly integrated and, secondly, that the decision to make one shoot type or another is tied to the resource status of the whole rhizome system, rather than to that portion of the system containing the differentiating bud. Studies of the translocation of ^{14}C-assimilate confirm that the rhizome systems of mayapple are highly integrated with respect to photosynthetic assimilate (Watson, 1990; Landa et al., 1992). Although the details are only now being revealed, it seems clear that in this plant the interaction between developmental phenology and environmental variability modulates the effects of local environmental variation, but it does so at the expense of maintaining the plant's capacity to respond plastically throughout the growing season. This may constrain the potential for rapid response to environmental change, but ensures that the plant does not use misleading local parameters for making demographically important developmental decisions (Watson, 1990).

Harvest index and crop yield

The harvested sink

The allocation of resources to reproductive growth and development is of fundamental importance in determining the yield potential of seed crops. Reproductive sinks are large; in modern cultivars of grain legumes and cereals, a harvest index (HI) of around 50% is achieved in favourable growing conditions. For example, in temperate cereals the ear makes up around 50% of the final biomass of the main shoot (Fig. 1b); this reflects both the establishment of a large number of reproductive sites, i.e. fertile florets, and the preferential partitioning of assimilate from the ear itself and especially from the upper leaves of the shoot to the grains throughout the period of grain growth. The developing grains thus tend to monopolize the supply of current assimilate. In addition grain growth may also be supported by the mobilization of reserve carbohydrates, mainly from the internodes of the stem, that are accumulated in both the pre- and the post-anthesis periods. It is estimated that pre-anthesis assimilate may contribute up to about 10% of the final biomass of the grain in unstressed field conditions, but under conditions of high temperature and water stress during grain-filling this contribution may rise to over 40% (Austin *et al.*, 1977*b*, 1980*a*).

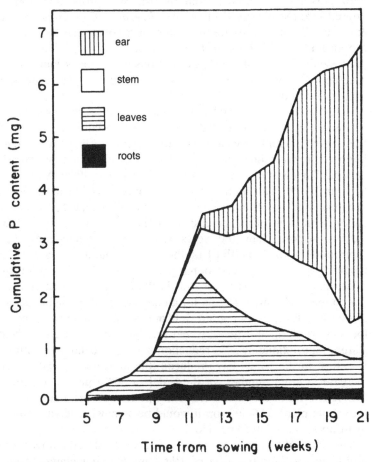

Fig. 4. Cumulative phosphorus content of the main shoot and root system through the life cycle of wheat (after Mohamed & Marshall, 1979).

The cereal ear is also a large sink for nitrogen and phosphorus (McNeal, Berg & Watson, 1966; Mohamed & Marshall, 1979). For example, in a field study of spring barley it was found that the main shoot ear accumulated 80% of the total phosphorus of the shoot (Fig. 4) and this was associated with the very high P content of the grain (Mohamed & Marshall, 1979). In this study, approximately two-thirds of the total ear P was obtained by root uptake from the soil from anthesis onwards and the remainder was derived from internal sources, i.e. by re-translocation from leaf and stem tissues. Similarly, in wheat, about half the N in the ear may normally be derived from re-translocation, but under conditions of poor post-anthesis uptake from the soil this proportion may rise to around 75% (McNeal et al., 1966; Austin et al., 1977a; Gregory, Marshall & Biscoe, 1981). Leaves are the major source of re-translocated N; as the proportion of N exported from an individual leaf increases so its photosynthetic capacity declines, and this decline is accompanied by its senescence. Thus the gain of N by the grains is at the expense of the supply of carbon.

In temperate cereals and legumes it appears that the very large increase (of the order of 60%) in seed yield of modern cultivars compared with their ancestors and very early varieties is more or less entirely due to an improved harvest index, as the total above-ground biomass production is very similar (Austin et al., 1980b; Gifford et al., 1984; Austin, Ford & Morgan, 1989). This improvement in economic yield can be viewed as a progressive trade-off between vegetative and reproductive resource allocation with time in an increasingly uniform environment. In very old varieties (more than 100 years since their introduction) the grain represents around one-third of the above-ground biomass whereas in modern varieties this proportion is just over 50% (Table 1) (Austin, Ford & Morgan, 1989). This major shift in the HI of temperate cereals is not due to the deliberate selection for an improved HI per se but is the result of selection for increased grain yield and for shorter straw to reduce crop lodging. Modern cultivars of wheat are thus characterized by their short stature (approximately 50% reduction in height) compared with very old varieties; further, they have reduced tillering capacity and a significant increase in grain number (of the order of 30%) per ear (Austin et al., 1989). The HI of the main shoot ear is commonly just over 50%, but for the whole plant or field crop it would be somewhat lower owing to the inclusion of lower-yielding tillers.

The reduced height of the most recently introduced cultivars of wheat is due to the incorporation of Rht semi-dwarfing genes by the plant breeder. An examination of the differences in the pattern of dry weight allocation between the developing ear and stem in closely related semi-

Table 1. *Yield components and HI of winter wheat varieties as group means based on time since introduction*

Age group	No. of varieties	Grain (g m^{-2})	Straw (g m^{-2})	Biomass (g m^{-2})	HI
Very old (1830–1907)	4	505	995	1500	0.34
Old (1908–16)	2	557	983	1541	0.36
Intermediate (1953–72)	2	669	814	1484	0.45
Modern (1981–6)	5	805	784	1588	0.51

Source: After Austin *et al.* (1989).

dwarf (*Rht2*) and tall (*rht*) genotypes of wheat indicates that in the former a far greater proportion of the pre-anthesis dry matter is partitioned to the ear than to the stem (Brooking & Kirby, 1981). This results in a greater ear mass at anthesis and the production of more fertile florets per spikelet, which in turn leads to more grains set per spikelet and per ear, a higher grain yield per shoot and a greater HI. Similar observations have been reported recently by Siddique, Kirby & Perry (1989) in a detailed survey of ear : stem ratios in a range of Australian wheat cultivars and in several near-isogenic lines differing only in *Rht* dwarfing genes. The improvement in grain yield, via the production of more grains per ear, associated with the introduction of dwarfing genes thus appears to be closely related to the change in competitive relations between the developing ear and the stem, with the latter competing less strongly than the ear for assimilate. Surprisingly, this pattern of development does not seem to be directly associated with the reduced growth potential of the stem but is more related to some independent activity, as yet undefined, of the *Rht* gene. Nevertheless it can be proposed that deliberate selection for high ear : stem dry mass ratios at anthesis may result in some further improvement in the yield potential of temperate cereals.

Further improvement in harvest index

The attainment of higher productivity in the future by further improvement in the HI of cereals such as wheat seems to be limited unless there is also a significant increase in the acquisition of resources. Although, in current cultivars, there is considerable sink potential that remains to be realized by improvement in grain set and increase in grain size, further resources will be required to support additional sink growth. This will be particularly important in the immediate post-anthesis period during the critical stages of grain set and early grain growth that determine both grain number per spikelet and the number of endosperm cells per grain (which is very closely correlated with final grain size) (Hay & Walker, 1989). Thus any major additional shift in the balance between vegetative and reproductive growth may well become counter-productive as a point must eventually be reached where reduced investment in leaves and stems will lead to less efficient light interception by the crop canopy. As a result, current assimilation, levels of reserve carbohydrates, and the mineral nutrient status of vegetative biomass that will be necessary to support the increased requirements of the larger reproductive sink, will in due course become growth-limiting. Nonetheless, Austin *et al.* (1980*b*) have proposed that a winter wheat with a HI of 0.62 would be theoretically possible if half of the stem and leaf-sheath biomass of a modern cultivar could be re-partitioned to the ear sink. This would result in an increase in grain yield of about 26% over current genotypes. The viability of such a prospect remains to be explored.

However, the principle of improvement in HI at the expense of vegetative structures is a central feature of Donald's (1968) proposal for a cereal ideotype and of ideotypes for other annual seed crops (Donald & Hamblin, 1983). It is proposed that the competitive ability of an individual, which is important for its growth and survival in mixed communities, is of far less significance in pure stands (of single genotype). Correspondingly in the latter situation it is argued that poor competitors will be the most successful as they will interfere only minimally with each other's growth. Thus reductions in vegetative growth, i.e. in plant height, leafiness, and degree of branching or tillering, all of which are associated with competitiveness, are prescribed for the crop ideotype; if reproductive growth is able to be sustained, then a plant with a high harvest index will result. Plants of this form would be the ideal communal plants but as they would tend to have quite low individual plant yields, owing to the loss of tillers or branches, they would have to be sown at high density in order to exploit the environment and so express their yield potential per unit area of land (Donald & Hamblin, 1976; Donald, 1981). The communal plant can thus be viewed as the product of selection for 'unfitness' but when grown at high density in the relatively stable arable agroecosystem its large reproductive sink and reduced vegetative structures are likely to maximize the investment of resources in seed yield.

Although there is little information on the performance of the communal ideotype, as very few appropriate genotypes exist, limited observations on semi-dwarf uniculm barley selections with a high HI indicate that they have the potential to significantly outyield established tillering cultivars in the field (Donald, 1979). The mean HI of these selections is close to 0.5 with a maximum value of 0.54. In wheat, uniculm selections with 'gigas' characters (large, thick leaves, a tall stem, and a large ear with many grains) have a HI of 0.55 in the field at wide spacing (Atsmon & Jacobs, 1977), but low tillering lines derived from this material, although very productive, have a HI of only 0.39 in field plots; this value is very similar to that for a standard-tillering cultivar grown under the same conditions (Marshall & Boyd, 1985). However, in a well-watered glasshouse environment these lines have HI values approaching 0.5 (C. Marshall, unpublished). More recently Richards (1988) has described the growth of new low tillering lines derived from a 'gigas' uniculm selection. Low tillering was associated with a high HI, the uniculm phenotypes having a HI of 0.47 under glasshouse conditions.

These examples clearly show that, despite selection for restricted tillering capacity (a major feature of the proposed cereal ideotype), there has not as yet been a significant improvement in HI over standard cultivars. This could perhaps be realized by the addition of *Rht* dwarfing genes to existing uniculm and low-tillering lines or by the chemical manipulation of stem growth of these lines by means of plant growth regulators. However, the latter approach may be somewhat counter-productive as such treatments commonly result in a stimulation of tillering, although tiller survival tends to be poor (Woodward & Marshall, 1989).

Modification of harvest index, via a relative reduction in vegetative structures, is an important strategy for improving crop yields in environments limited by water supply. In wheat in dryland conditions, Islam & Sedgley (1981) found that de-tillering a tillering cultivar to just two shoots per plant (a biculm phenotype), thereby attaining a constant shoot density per unit area, gave an increase in grain yield of up to 22% over a freely tillering crop (Table 2). This was associated with a significant improvement in HI (from 0.24 to 0.30) and a more conservative pattern of water use by the crop owing to the reduction in canopy leaf area. As a result more water was available during the later stages of growth to sustain grain set and grain-filling. Similar observations have been made by Richards (1983) and Richards & Townley-Smith (1987) in the glasshouse for de-tillered and partly defoliated plants and for a range of genotypes differing in tillering capacity. Similary, restricting plants of chickpea in the field to only two major branches significantly increased seed yield over freely branching control plants (Siddique & Sedgley, 1985). The increased yield was again associated with improved water use, which in turn allowed a greater expression of reproductive potential via an improvement in HI of the order of 30% (Table 2).

Table 2. *Effect of de-tillering and de-branching on yield and HI of wheat and chickpea*

	Grain yield (t ha^{-1})	HI
Wheat		
Control	1.80	0.35
De-tillered	2.05	0.39
Chickpea		
Control	1.35	0.26
De-branched	1.87	0.34

Source: After Islam & Sedgley (1981) and Siddique & Sedgley (1985).

These results clearly demonstrate that the loss of structures associated with competitive ability can lead to a greater overall investment of resources in reproductive growth where water is a limiting resource. It should therefore be possible to increase HI and hence seed yield of crops such as wheat and chickpea that are grown in relatively dry environments by selection for restricted tillering and branching. In effect this represents selection for reduction in the overall size of the vegetative sink as formulated for the annual seed crop ideotype (Donald & Hamblin, 1983).

Conclusions

All kinds of plant biologists, from the theoretical to the very applied, have contributed to our understanding of various aspects of reproductive allocation in plants. It is therefore perhaps not so surprising that there is such a diverse array of information on this topic. The traditional approach, based on the presumed strategic allocation of limiting resources with trade-off between different functional activities, has been valuable in focusing on the variation in the patterns of allocation of different resources such as energy or phosphorus between species and between contrasting environments. However, too often such studies have been concerned with single rather than multiple resources, and the extent to which such factors exert a growth-limiting role is far from clear. This situation is further confounded by the high degree of carbon or energy independence of many reproductive structures and by the extensive recycling of nutrients from vegetative to reproductive structures. The interpretation and general extrapolation of results, especially in functional terms, is therefore far from precise. Furthermore, it is clear that there are many examples where the pattern of plant development *per se*, via meristematic activity and module production, determines the way that resources become allocated. Thus the inherent developmental programme of an individual that regulates the production, type (i.e. vegetative or reproductive) and growth potential of meristems is of fundamental significance in determining the nature and degree of response to fluctuating resources. Indeed there appear to be situations where the availability of meristems – rather than the availability of finite resources – is the key factor limiting growth potential. This is clearly an area that requires further detailed study. Analysis at the morphological level thus provides an alternative perspective on resource allocation and its significance should not be overlooked in traditional resource-based studies of allocation.

At the agronomic level, it is clear that increases in seed yield of annual crops are due to changes in reproductive allocation rather than to increases in the rate of leaf photosynthesis. Selection and plant breeding, and also experimental source–sink manipulations, have thus improved harvest index, but there seems relatively little scope for further improvement, especially in resource-rich environments, unless additional crop biomass can be produced by enhanced photosynthesis. On the other hand in less favourable situations, for example where growth and yield are restricted by water supply, there is a real prospect that some reduction in vegetative structures may increase yield via an improved harvest index that is associated with a more conservative use of water. Thus sink potential will be able to be more fully expressed. However, the general prospect of significantly increasing yield by directly improving the strength or size of reproductive sinks, despite some spare capacity, seems unlikely to be realized.

Acknowledgement

Part of this work was supported by a National Science Foundation grant to M.A.W.

References

Abrahamson, W.G. (1979). Patterns of resource allocation in wildflower populations of fields and woods. *American Journal of Botany* **66**, 71–9.

Abrahamson, W.G. & Caswell, H. (1982). On the comparative allocation of biomass, energy and nutrients in plants. *Ecology* **63**, 982–91.

Abrahamson, W.G. & Gadgil, M. (1973). Growth form and reproductive effort in goldenrods (*Solidago*, Compositae). *American Naturalist* **107**, 651–61.

Atsmon, D. & Jacobs, E. (1977). A newly bred 'Gigas' form of bread wheat (*Triticum aestivum* L.): morphological features and thermo-photoperiodic responses. *Crop Science* **17**, 31–5.

Austin, R.B., Bingham, J., Blackwell, R.D., Evans, L.T., Ford, M.A., Morgan, C.L. & Taylor, M. (1980*b*). Genetic improvements in winter wheat yields since 1900 and associated physiological changes. *Journal of Agricultural Science, Cambridge* **94**, 675–89.

Austin, R.B., Edrich, J.A., Ford, M.A. & Blackwell, R.D. (1977*b*). The fate of the dry matter, carbohydrates and ¹⁴C lost from the leaves and stems of wheat during grain filling. *Annals of Botany* **41**, 1309–21.

Austin, R.B., Ford, M.A., Edrich, J.A. & Blackwell, R.D. (1977*a*). The nitrogen economy of winter wheat. *Journal of Agricultural Science, Cambridge* **88**, 159–67.

Austin, R.B., Ford, M.A. & Morgan, C.L. (1989). Genetic improvement in the yield of winter wheat: a further evaluation. *Journal of Agricultural Science, Cambridge* **112**, 295–301.

Austin, R.B., Morgan, C.L., Ford, M.A. & Blackwell, R.D. (1980*a*). Contributions to grain yield from pre-anthesis assimilation in tall and dwarf barley phenotypes in two contrasting seasons. *Annals of Botany* **45**, 309–19.

Bangerth, F. (1989). Dominance among fruit/sinks and the search for a correlative signal. *Physiologia Plantarum* **76**, 608–14.

Bazzaz, F.A. & Carlson, R.W. (1979). Photosynthetic contribution of flowers and seeds to reproductive effort of an annual colonizer. *New Phytologist* **82**, 223–32.

Bazzaz, F.A., Carlson, R.W. & Harper, J.L. (1979). Contribution to the reproductive effort by photosynthesis of flowers and fruits. *Nature* **279**, 554–5.

Bazzaz, F.A., Chiariello, N.R., Coley, P.D. & Pitelka, L.F. (1987). Allocating resources to reproduction and defense. *Bioscience* **37**, 58–67.

Bazzaz, F.A. & Reekie, E.G. (1985). The meaning and measurement of reproductive effort in plants. In *Studies in Plant Demography: A Festschrift for John L. Harper* (ed. J. White), pp. 373–87. London: Academic Press.

Benner, B.L. & Watson, M.A. (1989). Developmental ecology of mayapple: seasonal patterns of resource distribution in sexual and vegetative rhizome systems. *Functional Ecology* **3**, 539–47.

Biscoe, P.V., Gallagher, J.N., Littleton, E.J., Monteith, J.L. & Scott, R.K. (1975). Barley and its environment. IV. Sources of assimilate for the grain. *Journal of Applied Ecology* **12**, 295–318.

Brenner, M.L. (1988). The role of hormones in photosynthate partitioning and seed filling. In *Plant Hormones and their Role in Plant Growth and Development* (ed. P.J. Davies), pp. 474–93. Dordrecht: Kluwer.

Brooking, I.R. & Kirby, E.J.M. (1981). Interrelationships between stem and ear development in winter wheat: the effects of a Norin 10 dwarfing gene, Gai/Rht. *Journal of Agricultural Science, Cambridge* **97**, 373–81.

Chan, B.G. & Cain, J.C. (1967). The effect of seed formation on subsequent flowering in apple. *Proceedings of the American Society for Horticultural Science* **91**, 63–8.

Clifford, P.E. (1977). Tiller bud suppression in reproductive plants of *Lolium multiflorum* Lam. cv. Westerwoldicum. *Annals of Botany* **41**, 605–15.

Cody, M. (1966). A general theory of clutch size. *Evolution* **20**, 174–84.

Davy, A.J. & Smith, H. (1988). Life-history variation and environment. In *Plant Population Ecology* (ed. A.J. Davy, M.J. Hutchings & A.R. Watkinson), pp. 1–22. Oxford: Blackwell.

Donald, C.M. (1968). The breeding of crop ideotypes. *Euphytica* **17**, 385–403.

Donald, C.M. (1979). A barley breeding programme based on an ideotype. *Journal of Agricultural Science, Cambridge* **93**, 261–9.

Donald, C.M. (1981). Competitive plants, communal plants, and yield in wheat crops. In *Wheat Science – Today and Tomorrow* (ed. L.T. Evans & W.J. Peacock), pp. 223–47. Cambridge University Press.

Donald, C.M. & Hamblin, J. (1976). The biological yield and harvest index of cereals as agronomic and plant breeding criteria. *Advances in Agronomy* **28**, 361–405.

Donald, C.M. & Hamblin, J. (1983). The convergent evolution of annual seed crops in agriculture. *Advances in Agronomy* **36**, 97–143.

Eis, S., Garman, E.H. & Ebel, L.F. (1965). Relation between cone production and diameter increment of douglas fir (*Pseudotsuga menziesii* (Mirb) Franco), grand fir (*Abies grandis* Dougl.) and western white pine (*Pinus monticola* Dougl.). *Canadian Journal of Botany* **43**, 1553–9.

Evans, L.T. & Wardlaw, I.F. (1976). Aspects of the comparative physiology of grain yield in cereals. *Advances in Agronomy* **28**, 301–59.

Fischer, K.S. & Wilson, G.L. (1975). Studies of grain production in *Sorghum bicolor* (L. Moench). III. The relative importance of assimilate supply, grain growth capacity and transport system. *Australian Journal of Agricultural Research* **26**, 11–23.

Fitter, A.H. (1986). Acquisition and utilization of resources. In *Plant Ecology* (ed. M. Crawley), pp. 375–405. Oxford: Blackwell.

Fitter, A.H. & Setters, N.L. (1988). Vegetative and reproductive allocation of phosphorus and potassium in relation to biomass in six species of *Viola*. *Journal of Ecology* **76**, 617–36.

Flinn, A.M. & Pate, J.S. (1970). A quantitative study of carbon transfer from pod and subtending leaf to the ripening seeds of the field pea (*Pisum arvense* L.). *Journal of Experimental Botany* **21**, 71–82.

Foerste, A.F. (1891). On the formation of flower buds of spring-blooming plants during the preceding summer. *Bulletin of the Torrey Botanical Club* **11**, 62–4.

Geber, M.A. (1989). The interplay of morphology and development on size inequality: a *Polygonum* greenhouse study. *Ecological Monographs* **59**, 267–88.

Geber, M.A. (1990). The costs of meristem limitation in *Polygonum arenasterum*: negative genetic correlations between fecundity and growth. *Evolution* **44**, 799–819.

Geber, M.A., Watson, M.A. & Furnish, R. (1992). Genetic differences in clonal demography in water hyacinth *Journal of Ecology* (in press).

Gifford, R.M., Thorne, J.H., Hitz, W.D. & Giaquinta, R.T. (1984). Crop productivity and photoassimilate partitioning. *Science* **225**, 801–8.

Gregory, P.J., Marshall, B. & Biscoe, P.V. (1981). Nutrient relations of winter wheat. 3. Nitrogen uptake, photosynthesis of flag leaves and translocation of nitrogen to the grain. *Journal of Agricultural Science, Cambridge* **96**, 539–47.

Grime, J.P. (1979). *Plant Strategies and Vegetation Processes*. Chichester: Wiley.

Grime, J.P., Crick, J.C. & Rincon, J.E. (1986). The ecological significance of plasticity. In *Plasticity in Plants* (ed. D.H. Jennings & A.J. Trewavas), pp. 5–29. Cambridge: Company of Biologists.

Harper, J.L. (1977). *The Population Biology of Plants*. London: Academic Press.

Harper, J.L. & Ogden, J. (1970). The reproductive strategy of higher plants. I. The concept of strategy with special reference to *Senecio vulgaris* L. *Journal of Ecology* **58**, 681–98.

Harrison, M.A. & Kaufman, P.B. (1980). Hormonal regulation of lateral bud (tiller) release in oats (*Avena sativa* L.) *Plant Physiology* **66**, 1123–7.

Hay, R.K.M. & Walker, A.J. (1989). *An Introduction to the Physiology of Crop Yield*. Harlow: Longman.

Hickman, J.C. (1975). Environmental unpredictability and plastic energy allocation strategies in the annual *Polygonum cascadense* (Polygonaceae). *Journal of Ecology* **63**, 689–701.

Hickman, J.C. & Pitelka, L.F. (1975). Dry weight indicates energy allocation in ecological strategy analysis of plants. *Oecologia* **21**, 117–21.

Holm, T. (1899). *Podophyllum peltatum*, a morphological study. *Botanical Gazette* **27**, 419–33.

Islam, T.M.T. & Sedgley, R.H. (1981). Evidence for a 'uniculm effect' in spring wheat (*Triticum aestivum* L.) in a Mediterranean environment. *Euphytica* **30**, 277–82.

Jewiss, O.R. (1972). Tillering in grasses – its significance and control. *Journal of the British Grassland Society* **27**, 65–82.

Landa, K., Benner, B., Watson, M.A. & Gartner, J. (1992). Physiological integration for carbon in mayapple (*Podophyllum peltatum*), a clonal perennial herb. *Oikos* (in press).

Law, R. (1979). The costs of reproduction in an annual meadow grass. *American Naturalist* **113**, 3–16.

Law, R., Bradshaw, A.D. & Putwain, P.D. (1977). Life history variation in *Poa annua*. *Evolution* **31**, 233–46.

Lovett Doust, J. (1980). Experimental manipulation of patterns of resource allocation in the growth cycle and reproduction of *Smyrnium olusatrum* L. *Biological Journal of the Linnean Society* **13**, 155–66.

McNeal, F.H., Berg, M.A. & Watson, C.A. (1966). Nitrogen and dry matter in five spring wheat varieties at successive stages of development. *Agronomy Journal* **58**, 605–8.

Maillette, L. (1982). Structural dynamics of silver birch. 1. The fates of buds. *Journal of Applied Ecology* **19**, 203–18.

Marshall, C. & Boyd, W.J.R. (1985). A comparison of the growth and development of biculm wheat lines with freely tillering cultivars. *Journal of Agricultural Science, Cambridge* **104**, 163–71.

Marshall, C. & Ludlam, D. (1989). The pattern of abortion of developing seeds in *Lolium perenne* L. *Annals of Botany* **63**, 19–27.

Mohamed, G.E.S. & Marshall, C. (1979). The pattern of distribution of phosphorus and dry matter with time in spring wheat. *Annals of Botany* **44**, 721–30.

Nooden, L.D. & Guiamet, J.J. (1989). Regulation of assimilation and senescence by the fruit in monocarpic plants. *Physiologia Plantarum* **77**, 267–74.

Ogden, J. (1974). The reproductive strategy of higher plants. II. The reproductive strategy of *Tussilago farfara* L. *Journal of Ecology* **62**, 291–324.

Pate, J.S. (1984). The carbon and nitrogen nutrition of fruit and seed – case studies of selected grain legumes. In *Seed Physiology*, volume 1: *Development* (ed. D.R. Murray), pp. 41–82. North Ryde, NSW: Academic Press, Australia.

Pitelka, L.F., Ashmun, J.W. & Brown, R.L. (1985). The relationship between seasonal variation in light intensity, ramet size, and sexual reproduction in natural and experimental populations of *Aster acuminatus* (Compositae). *American Journal of Botany* **72**, 311–19.

Pitelka, L.F., Stanton, D.S. & Peckenham, M.O. (1980). Effects of light and density on resource allocation in a forest herb, *Aster acuminatus* (Compositae). *American Journal of Botany* **67**, 942–8.

Primack, R.B. & Antonovics, J. (1982). Experimental ecological genetics in *Plantago*. VII. Reproductive effort in populations of *P. lanceolata* L. *Evolution* **36**, 742–52.

Richards, J. (1980). *The developmental basis of morphological plasticity in the water hyacinth*, Eichhornia crassipes *Solms*. Ph.D. dissertation, University of California, Berkeley.

Richards, R.A. (1983). Manipulation of leaf area and its effect on grain yield in droughted wheat. *Australian Journal of Agricultural Research* **34**, 23–31.

Richards, R.A. (1988). A tiller inhibitor gene in wheat and its effect on plant growth. *Australian Journal of Agricultural Research* **39**, 749–57.

Richards, R.A. & Townley-Smith, T.F. (1987). Variation in leaf area development and its effect on water use, yield and harvest index of droughted wheat. *Australian Journal of Agricultural Research* **38**, 983–92.

Siddique, K.H.M., Kirby, E.J.M. & Perry, M.W. (1989). Ear: stem ratio in old and modern wheat varieties; relationship with improvement in number of grains per ear and yield. *Field Crops Research* **21**, 59–78.

Siddique, K.H.M. & Sedgley, R.H.M. (1985). The effect of reduced branching on yield and water use of chickpea (*Cicer arietinum* L.) in a Mediterranean type environment. *Field Crops Research* **12**, 251–69.

Sinclair, T.R. & de Wit, C.T. (1975). Photosynthate and nitrogen requirements for seed production by various crops. *Science* **189**, 565–7.

Sohn, J.J. & Policansky, D. (1977). The costs of reproduction in mayapple *Podophyllum peltatum* (Berberidaceae). *Ecology* **58**, 1366–74.

Stephenson, A.G. (1981). Flower and fruit abortion: proximate causes and ultimate functions. *Annual Review of Ecology and Systematics* **12**, 253–79.

Stephenson, A.G., Devlin, B. & Horton, J.B. (1988). The effects of seed number and prior fruit dominance on the patterns of fruit production in *Cucurbita pepo* (zucchini squash). *Annals of Botany* **62**, 653–61.

Tamas, I.A., Engels, C.J., Kaplan, S.L., Ozbun, J.L. & Wallace, D.H. (1981). Role of indoleacetic acid and abscisic acid in the correlative control by fruits of axillary bud development and leaf senescence. *Plant Physiology* **68**, 476–81.

Tamas, I.A., Ozbun, J.L., Wallace, D.H., Powell, L.E. & Engels, C.J. (1979a). Effect of fruits on dormancy and abscisic acid concentration in the axillary buds of *Phaseolus vulgaris* L. *Plant Physiology* **64**, 615–19.

Tamas, I.A., Wallace, D.H., Ludford, P.M. & Ozbun, J.L. (1979b). Effect of older fruits on abortion and abscisic acid concentration of younger fruits in *Phaseolus vulgaris* L. *Plant Physiology* **64**, 620–2.

Thompson, K. & Stewart, A.J.A. (1981). The measurement and meaning of reproductive effort in plants. *American Naturalist* **117**, 205–11.

Tuomi, J., Makala, T. & Haukioja, E. (1983). Alternative concepts of reproductive effort, costs of reproduction, and selection in life-history evolution. *American Zoologist* **23**, 25–34.

Tuomi, J., Niemela, P. & Mannila, R. (1982). Resource allocation on dwarf shoots of birch (*Betula pendula*): reproduction and leaf growth. *New Phytologist* **91**, 483–7.

Van Andel, J. & Vera, F. (1977). Reproductive allocation in *Senecio sylvaticus* and *Chamaenerion angustifolium* in relation to mineral nutrition. *Journal of Ecology* **65**, 747–58.

Waite, S. & Hutchings, M.J. (1982). Plastic energy allocation patterns in *Plantago coronopus*. *Oikos* **38**, 333–42.

Wareing, P.F. & Phillips, I.D.J. (1981). *Growth and Differentiation in Plants* (3rd edition). Oxford: Pergamon Press.

Warren Wilson, J. (1972). Control of crop processes. In *Crop Processes in Controlled Environments* (ed. A.R. Rees, K.E. Cockshull, D.W. Hand & R.G. Hurd), pp. 7–30. London: Academic Press.

Watson, M.A. (1984). Developmental constraints: Effect on population growth and patterns of resource allocation in a clonal plant. *American Naturalist* **123**, 411–26.

Watson, M.A. (1990). Phenological effects on clone development and demography. In *Clonal Growth in Plants: Regulation and Function* (ed. J. van Groenendael & H. de Kroon), pp. 43–55. The Hague: SPB Press.

Watson, M.A. & Brochier, J. (1988). The role of changing nutrient levels on inflorescence induction in water hyacinth. *Aquatic Botany* **31**, 367–72.

Watson, M.A., Carrier, J.C. & Cook, G.S. (1982). Effect of exogenously supplied gibberellic acid (GA$_{23}$) on patterns of water hyacinth development. *Aquatic Botany* **13**, 57–68.

Watson, M.A. & Casper, B.B. (1984). Morphogenetic constraints on patterns on carbon distribution in plants. *Annual Review of Ecology and Systematics* **15**, 233–58.

Watson, M.A. & Cook, C. (1982). The development of spatial pattern in clones of an aquatic plant, *Eichhornia crassipes* Solms. *American Journal of Botany* **69**, 248–53.

Watson, M.A. & Cook, G.S. (1987). Demographic and developmental differences among clones of water hyacinth. *Journal of Ecology* **75**, 439–57.

White, J. (1979). The plant as a metapopulation. *Annual Review of Ecology and Systematics* **10**, 109–45.

White, J. (1984). Plant metamerism. In *Perspectives in Plant Population Ecology* (ed. R. Dirzo & J. Sarukhan), pp. 15–47. Sunderland: Sinauer.

Wiens, D. (1984). Ovule survivorship, brood size, life history, breeding systems and reproductive success in plants. *Oecologia* **64**, 47–53.

Wiens, D., Calvin, C.L., Wilson, C.A., Davern, C.I., Frank, D. &

Seavey, S.R. (1987). Reproductive success, spontaneous embryo abortion and genetic load in flowering plants. *Oecologia* **71**, 501–9.

Woodward, E.J. & Marshall, C. (1989). Effects of plant growth regulators on tiller bud outgrowth in uniculm cereals. *Annals of Applied Biology* **114**, 597–608.

C.D. PIGOTT

Are the distributions of species determined by failure to set seed?

Introduction

It is a widely accepted principle of plant geography that the distribution of vegetation and of plant species is primarily controlled by climate, but that climate changes with time (Cain, 1944). The basis of this principle is the broad correlation of vegetation with latitude, essentially a correlation with temperature, and the modification of this relationship by the availability of water.

These correlations are well illustrated by the boundaries of the climax woodlands of northern Europe, which run latitudinally across northern Russia and Finland but then have a southwestern trend across Scandinavia and Britain (Sjörs, 1965). These boundaries are correlated with various measures of summer warmth. The northern limits of many of the constituent species are similarly correlated with summer warmth but also with the vegetational zones themselves.

The possibility that temperature controls the boundaries through its effect on reproduction rather than vegetative growth is suggested from the common observation that heavy crops of fertile fruit of some of the dominant species of tree follow, or coincide with, years of exceptional warmth (Matthews, 1955). There is, however, very little information about the production of fertile fruit at the northern limits of species, nor are there many studies of the population dynamics of species at their limits, so that the relation between reproductive capacity and regeneration is unknown.

Studies of species in northwestern England

Recent studies on the distribution of vegetation in Britain for the National Vegetation Classification (Rodwell, 1991) provide many examples of plant communities that are restricted either to the warmer and drier southeast or to the cooler and wetter northwest of Britain. In a system based on species composition, this is partly a direct consequence of the many species that themselves have geographical limits with the same diagonal trend. Several of these species have their northern limit in northwestern England; some have been studied analytically and experimentally over a period of up to 25 years.

Tilia cordata (small-leaved lime)

The most detailed study has been of *Tilia cordata*. This tree, which was formerly a dominant species in the pre-settlement woodlands of the English lowlands, reaches a well-defined northern limit in the Lake District and Northumberland where it is fragmented into more than 150 populations (Fig. 1). Many of these populations are small and occupy cliffs or steep rocky slopes where they have escaped destruction by man. Their absence from slopes of northerly aspect is further evidence of a relation of the distribution of the tree to climate (Pigott & Huntley, 1978).

A remarkable feature of the populations is that they are largely composed of old trees, often of great size (Pigott, 1989), and there is no recent regeneration over a period of many years. Samples of fruit have

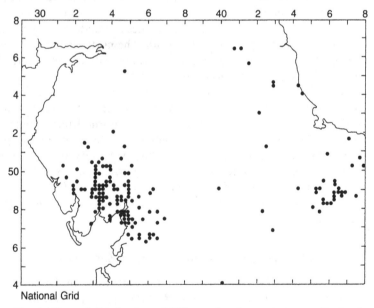

Fig. 1. Natural distribution of *Tilia cordata* in northwestern England, showing its northern limit in Britain (based on field records since 1964).

been collected almost every year from 1964 to 1989 from a population on a southerly aspect at Aughton, near Lancaster, and from other localities as opportunity has allowed. In most years the percentage of fertile fruit has been less than 1% even though samples collected in the same years from sites in the English Midlands or southern England have contained 10–50% of fertile fruit. Fruits that are not fertile contain either all ten ovules shrunken, or one incompletely developed seed.

Only in 1976, 1983 and 1984 have a significant proportion of the fruits in samples from northwestern England contained fully developed seeds. In these three years mean daily maxima of air temperatures at the time of flowering in late July and early August were 3–8 °C above the long-term average of about 19 °C, thus providing climatic conditions normally characteristic of southern England. Samples of styles collected from the flowers at 2–5 days after anthesis showed numerous pollen tubes extending the full length from the stigma to the ovules. Similar samples collected in normal or cooler years showed pollen tubes arrested, either in the superficial tissues of the stigma or at short distances down the style. An experimental study by Pigott & Huntley (1981) of the influence of temperature on growth of pollen tubes, both in sucrose solutions and in live styles, shows an unusually high temperature threshold of 15 °C for germination of the pollen and of 19–20 °C for rapid growth of the pollen tube (Fig. 2). There is therefore good evidence that fruit production is primarily controlled by temperature.

In all but two of the twenty-five years no seedlings were recorded at any northern site. This included 1977, following the warm summer of 1976 when the proportion of fertile fruits was 6–16%, which is approximately equivalent to a seeding density on the ground immediately around the parent trees of 10–20 m^{-2}. Even then no seedlings were found in several localities examined.

Fertility in 1983 and 1984 was much higher (64–86% at Rydal in the Lake District). Although seeding densities around these trees would be over 100–150 m^{-2}, the actual densities of seedlings in 1984 and 1985 were very low (0.03 m^{-2}). None survived beyond its first winter at any site. The seedlings are very palatable and were almost certainly destroyed by wood mice and voles, which also consume a high proportion of the fruits when these are sown experimentally at realistic densities (150 m^{-2}) over areas in woodland equivalent to that of a gap created by the loss of a single tree.

Successful regeneration in woodlands in the Midlands and southern England normally follows the occurrence of much higher densities of seedlings, which extend over large areas and are associated with a high density of parent trees. Then, in spite of high mortality from predation, a proportion of seedlings survive to become saplings. The only population

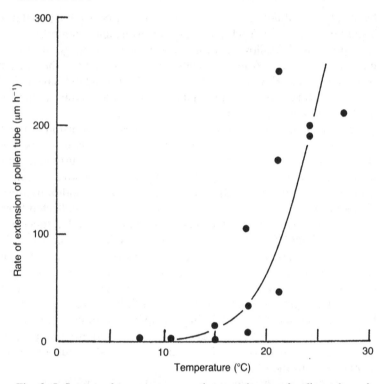

Fig. 2. Influence of temperature on the growth rate of pollen tubes of *Tilia cordata* in aqueous solution of sucrose (0.29 mol m^{-3} sucrose with 1.6×10^{-3} mol borate). From Pigott & Huntley (1981).

near the northern limit of *T. cordata* which includes saplings has a similar high proportion of parent trees in the canopy and occupies an exceptionally sheltered deep depression in the limestone of the sunny coastal region. The saplings all appear to have originated after the very warm summer of 1959 and there has been no regeneration since.

These observations strongly indicate that production of fertile fruit by *T. cordata* near the northern limit is normally inadequate to allow regeneration. Even after years as exceptional as 1983 and 1984 there was no regeneration; it seems that a much higher frequency of large crops of fertile fruit is needed to increase the probability of coincidence with, for example, gaps in the canopy in the right condition and low densities or predators. In most northern populations the density of parent trees is probably too low, partly because many of the old trees have been coppiced, so that the structure of the present woodlands is unfavourable for regeneration.

Tamus communis (black bryony)

The area within the British Isles occupied by *Tamus communis* is almost exactly the same as that of *Tilia cordata* but the two species have very different distributions in Europe. Whereas *T. cordata* is a Continental species, extending throughout the zone of summer-green deciduous woodlands northwards to Scandinavia and Finland and as far east as the Urals, *T. communis* has a very clearly defined Mediterranean–Atlantic distribution.

In northern England, the two species are most frequent in the same area, forming a crescent around Morecambe Bay (Fig. 3). This is a feature of the distribution of several other southern species including *Acer campestre*, *Cornus sanguinea* and *Rhamnus catharticus*. The actual limit of *Tamus* is sharply defined: it is frequent in hedges, certain types of woodland and wood margins in the coastal area and extends up each valley to a particular point at which it abruptly ceases.

Plants are dioecious but both male and female individuals usually occur together up to the limit. Fruits are produced prolifically in most years and there are on average three fertile seeds in each fruit. Samples of seed were collected regularly from 1968 to 1984 and almost all were found to be fertile.

The fruit are consumed in late autumn by blackbirds (*Turdus merula*), thrushes (*T. philomeles*), and starlings (*Sturnus vulgaris*); the hard seeds are passed undamaged. Dispersal is effective, as seedlings occur frequently under roost sites and have been observed along the base of a newly erected wire fence. In spite of successful regeneration within the area already occupied by the species and of production of fertile seed at the present limit, there is no evidence that *Tamus* is advancing northwards at the present day, either from the distribution of old records or from the occurrence of small outlying populations.

At several places along the northern limit in the Lake District and the eastern limit in this region at the foot of the Pennines, the boundary coincides with the limit of enclosure before the extension of enclosure in the early nineteenth century. At this boundary there is usually a change from old mixed hedgerows to either stone walls or new hedges composed largely of thorn (*Crataegus monogyna*). Although the absence from the boundaries of the new enclosures shows that *Tamus* has not advanced in the subsequent period of 170 years, the cause is confounded with the

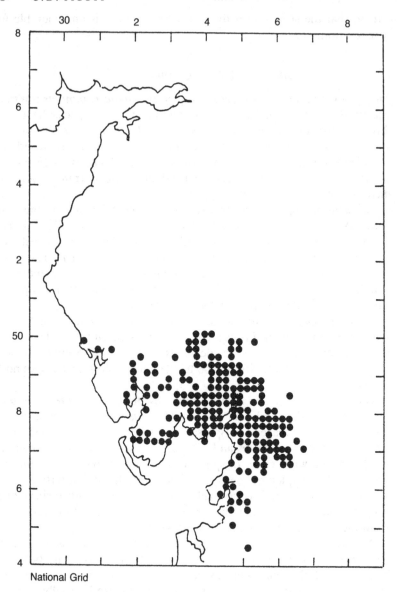

Fig. 3. Natural distribution of *Tamus communis* in northwestern England (based on field records since 1964).

change in habitat and sometimes with altitude. Plants do occur, however, along walls and in new hedges within the older settled landscape.

An experiment was conducted in 1979 to discover the conditions needed for establishment. Seeds were sown in October at three densities (10, 100 and 1000 seeds per 10 m length) along a mixed hedgerow, which had recently been laid, at a site situated close to but inside the natural limit. Seeds sown simultaneously in open ground produced seedlings and were almost all fertile, but no seedlings appeared in the experimental site either in 1980 or 1981 and no plants have become established since. The rarity of seedlings in and near hedges in relation to the abundance of fertile seed produced is described by Burkill (1944).

Tamus is an herbaceous climber; the stems die down each year but the diameter of the tuber increases at a rate of about 2 mm per year. Burkill (1944) estimates from the frequency distribution of size of tubers that the average plant is about 20 years old, but much older individuals also occur. A study of the population dynamics of the species at its northern limit is therefore possible and has now been started.

From the evidence available it would seem that production of fertile fruit is adequate to allow regeneration at the northern limit but only in particular conditions which have not yet been identified. Failure to spread northward is correlated with structural features of the habitat, as well as with climate.

Acer pseudoplatanus (sycamore)

This alien species provides an informative contrast with the native species discussed in the previous sections. Sycamore is native in the hills and mountains of central Europe (Fig. 4) where it is most characteristic of the altitudinal zones dominated by beech (*Fagus sylvatica*) and spruce (*Picea abies*) at 200–1500 m. The limits of its natural distribution are now not clearly defined but it has certainly spread westwards and northwards during the historic period. Although initially planted because of its valuable timber and ease of propagation, it has then spread spontaneously into woodlands. It is a common species throughout northwestern England, where it is at least 500 km northwest of its native limits, but it regenerates freely from seed in woodlands, especially those on moist but well-drained soils of moderate fertility.

Sycamore flowers in early May; even when temperatures are below average, the pollen germinates and the pollen tubes grow to the base of the style within 3–5 days of the flowers opening. An experimental study has shown that this can occur at temperatures of the flower as low as 5.5 °C (Pigott & Warr, 1989). Fruits containing fertilized ovules grow

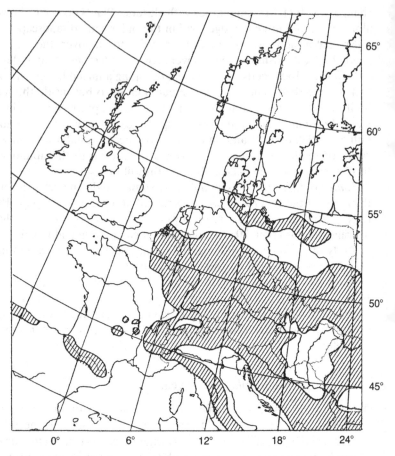

Fig. 4. Probable native distribution of *Acer pseudoplatanus* in western Europe (based on Meusel *et al.*, 1978).

throughout the summer, and although the size of crop varies considerably from year to year, there is no simple relationship with summer temperatures. For example, large crops of fertile fruit were produced in both the generally cold summer of 1980 and the exceptionally warm summer of 1984.

These results show that the production of fruit by sycamore is well adapted to the cool springs and summers of the mountains of central Europe, and this has allowed the species to exploit the oceanic climate of northern and western Britain. Sycamore is much less at home in eastern England, and in those areas where annual rainfall is less than 760 mm, the ability of the species to enter the canopy of woodlands is greatly reduced (Pigott, 1984). The area of natural distribution is unlikely to be determined by production of fertile fruit, but by rainfall and frequency of drought, which determine the success of the species when growing in competition in woodlands.

The dynamic nature of distribution limits

The three species discussed in previous sections have been selected to illustrate contrasting dynamics at their northern limits. In summary, the northern populations of *Tilia cordata* are composed predominantly of old individuals and there is almost no regeneration from seed. *Tamus communis* has a very similar northern limit in England, but produces large amounts of fertile seed. It regenerates within its limit but shows no evidence of advancing northwards. *Acer pseudoplatanus* is growing far beyond its natural limit but produces abundant seed, regenerates freely and is known to have extended its range, partly with human assistance and partly spontaneously.

The history of the distribution of *Tilia cordata* is known with unusual precision. Its pollen is specifically identifiable and not widely dispersed, so that its presence in small quantities (though not as isolated single grains), preserved in peats or lake muds, indicates the occurrence of flowering trees nearby. The past distribution of pollen (Fig. 5) shows that

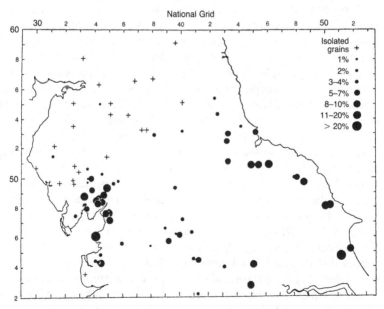

Fig. 5. Distribution of maximum values for the pollen of *Tilia* during the period from 5000 to 3000 BC (from Pigott & Huntley, 1980).

the species reached its present northern limit about 5000 years ago during the post-glacial (Flandrian) period of greatest summer warmth.

During that period summers are estimated to have been about 2 °C warmer than at present. This estimate was, in fact, originally based by Andersson (1902) on the difference between the past and present northern limit of hazel (*Corylus avellana*) in Scandinavia, in relation to the present summer isotherms. Unlike hazel, which retreated south as the summers became cooler, *Tilia cordata*, when not destroyed by man, survived right up to its former northernmost limit, where it is now in the true sense a relic of the earlier warm period. It survives because the individuals are essentially immortal: not only can the stems live for several centuries but, when they collapse, they are replaced by vigorous vegetative sprouts. Estimates of their age, based on various criteria, show that many of the trees are probably over a thousand years old (Pigott, 1989).

The present northern limit of *T. cordata* is bounded by the 18 °C isotherm for average daily maxima of air temperature (Fig. 6). When the limit was established this would have been the 20 °C isotherm, which now lies about 200 km further south. It defines an area in which the species produces fertile fruit much more regularly and regenerates in suitable types of woodland. Even in this species, therefore, the northern limit is not apparently caused by failure to produce fertile fruit.

To maintain itself without regeneration from seed depends on a species being able to survive vegetatively or even, as in some water plants and the woodland herb *Adoxa moschatellina*, being capable of dispersal in a vegetative state. *T. cordata* is not unique in this respect. For *Cornus sanguinea* all post-glacial records of its pollen and fruit-stones lie within the area it now occupies (Godwin, 1956). At its northern limit in the Pennines and Lake District it rarely produces fertile fruit and no seedlings have been observed, yet in southeastern England it produces large crops of fruit and regenerates freely, enabling it to invade abandoned pastures. Like *T. cordata* this shrub can maintain itself vegetatively and individual clones, though usually fragmented, may occupy very large areas.

The history of *Tamus communis* is not known, but there are other species with almost the same northern limit which respond similarly and produce frequent crops of fertile fruit. Both field maple (*Acer campestre*) and buckthorn (*Rhamnus catharticus*) are insect-pollinated, yet their pollen has been recorded in Scotland during the middle Flandrian. Their former distribution is not known in such detail as for *Tilia cordata* but it seems that they have retreated to their present northern limit in the Lake District, which is probably more or less in equilibrium with present climate. The populations of both species do not contain trees of great age and in some sites there is regeneration from seed.

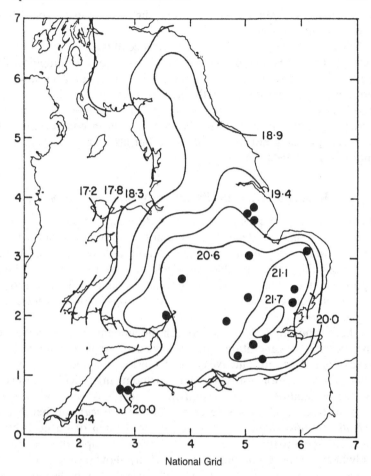

Fig. 6. Distribution of sites where frequent crops of fertile fruits of *Tilia cordata* have been recorded and where there is limited regeneration, in relation to the isotherms of average means of daily maxima of air temperature (reduced to sea level) during August for the period 1901–30 (based on Pigott & Huntley, 1981).

The spread northwards of sycamore (*Acer pseudoplatanus*) after its introduction into Britain, probably in the Middle Ages, is not essentially different from the history of some other introduced trees, such as sweet chestnut (*Castanea sativa*) and holm oak (*Quercus ilex*), or of the native species beech (*Fagus sylvatica*). All these trees have extended their range northwards during the historic period and have become integrated into characteristic plant communities, where they regenerate from seed. In all these species, unlike sycamore, the production of fertile fruit is correlated with summer warmth, either in the preceding year or in the year when the fruits are set and ripen.

These examples emphasize the importance of the history of a species in relation to understanding the control of its present geographical limits. Failure to produce fertile fruit is characteristic of relict species, which are no longer in equilibrium with present climate. Species that are in equilibrium are fertile at their northern limit but show no evidence of extending their range. Species that are not yet in equilibrium and are actively spreading not only fruit freely but also regenerate aggressively into existing vegetation.

Reproductive capacity and geographical limits

In none of the examples is failure to produce fertile fruit the cause of the geographical limit. Indeed, this could not be so simply because a species must be able to regenerate to reach its limit initially. There is, however, the possibility that production declines gradually as the limit is approached and the limit itself is the point along the climatic gradient at which the frequency and quantity of seed drops below that required for regeneration.

There is no simple relation between the amount of seed produced and the success of regeneration. Species have evolved as components of vegetation and their reproductive capacity must be adapted to the conditions which the vegetation normally offers, including in particular the structure of the vegetation and the competitive relations with other species. Certain trees, such as sessile oak (*Quercus petraea*) and beech (*Fagus sylvatica*), show marked periodicity in production of fertile fruit; they thus provide a natural experiment in varying the input of seed. In beech only heavy crops in full mast-years normally allow regeneration, and the sparse seedlings which arise from smaller crops entirely disappear (Watt, 1925). As the frequency of full mast-years decreases northwards, so does the success of regeneration.

The importance of the structure of the vegetation in allowing regeneration is well recognized silviculturally. In the uniform system, widely used in central France, in order to ensure successful regeneration from a heavy crop of acorns, about one-third of the parent trees must be cut. The aim is to achieve a density of oak seedlings of 50–80 m^{-2} ($0.5 - 0.8 \times 10^6$ ha^{-1}). It is then essential to reduce or eliminate competition from other faster-growing species, such as hornbeam (*Carpinus betulus*), or the oak suffers an unacceptable mortality from competition. Indeed, in a long-term study in which hornbeam and lime have been left undisturbed, oak, which was initially the most abundant seedling, has been virtually eliminated in 23 years. The frequency and size of mast crops of sessile oak (*Q. petraea*) decrease northwards; in Lorraine, this method of regeneration is no longer successful (Troup, 1966).

Although this is an artificial situation, it provides a clear indication of the types of observations and experiments that are needed to understand more natural systems. It also emphasizes the way in which human intervention in vegetation may alter the probability of successful regeneration. For some species, for example sycamore, present conditions are evidently very favourable: for other species, for example, small-leaved lime, it seems the reverse is true. The response of a species to climate is thus likely to be an interaction with other conditions. It is then not a simple matter to predict how a species will respond to climatic change, especially when its present distribution is not in equilibrium and was established in very different conditions from those that now prevail.

References

Andersson, G. (1902). Hasseln i Sverige fordom och nu. *Sveriges geologiska Undersökning. Afhandlingar och Uppsatser*, ser.C a, no. 3.

Burkill, I.H. (1944). Biological Flora of the British Isles. *Tamus communis* L. *Journal of Ecology* **32**, 121–9.

Cain, S.A. (1944). *Foundations of Plant Geography*. New York: Harper.

Godwin, H. (1956). *The History of the British Flora* (first edition). Cambridge University Press.

Matthews, J.D. (1955). The influence of weather on the frequency of beech mast years in England. *Forestry* **28**, 107–16.

Meusel, H., Jäger, E., Rauschert, S. & Weinert, E. (1978). *Vergleichende Chorologie der zentraleuropäischen Flora*. Jena: Fischer.

Pigott, C.D. (1984). The flora and vegetation of Britain: ecology and conservation. *New Phytologist* **98**, 119–28.

Pigott, C.D. (1989). Factors controlling the distribution of *Tilia cordata* Mill. at the northern boundary of its geographical range. 4. Estimated ages of the trees. *New Phytologist* **112**, 117–21.

Pigott, C.D. & Huntley, J.P. (1978). Factors controlling the distribution of *Tilia cordata* at the northern limits of its geographical range. 1. Distribution in north-west England. *New Phytologist* **81**, 429–41.

Pigott, C.D. & Huntley, J.P. (1980). Factors controlling the distribution of *Tilia cordata* at the northern limits of its geographical range. 2. History in north-west England. *New Phytologist* **84**, 145–64.

Pigott, C.D. & Huntley, J.P. (1981). Factors controlling the distribution of *Tilia cordata* at the northern limit of its geographical range. 3. Nature and causes of seed sterility. *New Phytologist* **87**, 817–39.

Pigott, C.D. & Warr, S.J. (1989). Pollination, fertilisation and fruit

development in sycamore (*Acer pseudoplatanus* L.). *New Phytologist*
111, 99–103.

Rodwell, J. (1991). *British Plant Communities. 1. Woodlands*. Cambridge University Press.

Sjörs, H. (1965). Forest regions. In *The Plant Cover of Sweden*. *Acta Phytogeographica Suecica* **50**, 48–63.

Troup, R.S. (1966). *Silvicultural Systems*. Oxford: Clarendon Press.

Watt, A.S. (1925). On the ecology of British beechwoods with special reference to their regeneration. *Journal of Ecology* **13**, 27–73.

Q.O.N. KAY

Edible fruits in a cool climate: the evolution and ecology of endozoochory in the European flora

Introduction

Seed plants show a diversity of adaptations for seed dispersal. These adaptations may involve physical dispersal agents (transport by wind, water, gravity, etc.) or biotic dispersal agents (self-dispersal, or more commonly transport by animals), often combined in various ways, simultaneously, sequentially or as alternatives. The adaptations that are possible or appropriate for a particular species are constrained by its structure and ecology, by its physical and biotic environment, by its history, and by genotypic limits to the extent to which it can vary in response to selective pressures. Adaptations for dispersal interact with and may be conditional upon adaptations for pollination (e.g. the position, number, structure and phenology of flowers) and for seedling establishment (seed size and number, seed protection, seed shape, seed dormancy and seed placement) (Fenner, 1985; Primack, 1987). Seed size has a particularly close interaction with seed dispersal mechanisms (Harper, Lovell & Moore, 1970).

Adaptations for seed dispersal must thus be seen in the context of the complex web of competing and interacting adaptations, constraints and possibilities that affect and ultimately determine the biology of a species. Within a species, the factors affecting seed dispersal may not be uniform. Different populations or individuals of a species may, for example, grow in widely different physical and biotic environments. Even within a single population, the biotic and physical environment may change greatly through time. A population of a pioneering tree species like *Rhamnus catharticus* or *Salix atrocinerea*, for example, may pass from a colonizing stage in an open habitat to a late-successional stage in closed woodland.

These uncertainties and complexities mean that the dispersal adaptations and mechanisms shown by a particular species or species population in a given geographic area may be far from optimal, especially in areas that have been disturbed in recent geological or historical time.

Much of the European flora has been drastically disturbed by cyclical climatic change and repeated glaciations during the Pleistocene period, and has subsequently been profoundly affected by expanding human populations and the impact of agriculture and other man-induced changes in the environment. Plants of climax communities are particularly vulnerable to such changes; their mechanisms for dispersal and establishment may have depended on balanced interactions or coevolved coadaptations with other species, and may fail or operate much less effectively in their absence. In contrast, many plants of open and colonizing communities, especially the weeds and ruderals that grow in man-made habitats, are well adapted to disturbed and unstable environments and show appropriate adaptations for dispersal and establishment.

Seed dispersal by frugivores

Edible fruits usually provide a reward for fruit-eating seed vectors in the form of nutritious fleshy pulp, juice or other edible material external to the seeds. In angiosperms the edible part of the fruit is most commonly formed by the ovary, but may consist of a fleshy aril or receptacle, or more rarely a fleshy perianth, bracts, or other structures. In modern gymnosperms it is formed by a fleshy aril or a fleshy cone. The seeds themselves are protected in various ways, usually either by a hard seed coat or by a stony endocarp (McKey, 1975). Strictly speaking, a fruit consists of the mature ovary and the seeds within it, but here, as in most discussions of the biology of fruit dispersal, the term will be used to describe the primary dispersal unit.

Seed dispersal in species with edible fruits may be divided into the essential *primary* stage, which is carried out by the frugivore after consumption of the fruit, and less clearly defined *secondary* (Matlack, 1989) and *tertiary* stages. The seeds are sometimes separated from the edible part of the fruit before they are swallowed, primary dispersal taking place when the rejected seeds are wiped off or discarded by the frugivore. More commonly they are consumed with the rest of the fruit, and primary dispersal takes place either when they are regurgitated with other indigestible material or when they are excreted. Primary dispersal frequently produces aggregations of seeds, which may be very large (Fenner, 1985). Secondary dispersal takes place when these aggregations are themselves dispersed, either by physical mechanisms (e.g. wind or water transport) or by biotic mechanisms (e.g. collection and dispersal or burial of seeds by ground-dwelling vertebrates, or by ants or other invertebrates, or incidental dispersal by birds, etc. feeding on food items in pellets or excreta, or even accidental transport by large herbivores). Tertiary dispersal is the process by which the seeds are emplaced in the soil; it normally depends on physical mechanisms and is related to the size, shape and surface characteristics of the seeds (Fenner, 1985). Little attention has been given to secondary and tertiary seed dispersal in studies of seed dispersal by frugivores.

Seed dispersal by frugivores (endozoochory; van der Pijl, 1982) has many parallels with pollen dispersal by flower-visiting invertebrate or vertebrate animals (zoophily) (Faegri & van der Pijl, 1979). In both cases a reward is usually provided for the animal vectors, associated with a distinctive and conspicuous plant organ, distinguished by colour, shape, position and often scent. Plant resources must be allocated to provide the reward, and for the distinguishing adaptations. Extensive coevolution and coadaptation have taken place between zoophilous flowers and pollinators, and between fruit-bearing plants and frugivorous animals, with varying degrees of specialization. The reward may be mainly in the form of energy-providing carbohydrate (nectar in flowers, sugar or starch, etc. in fruits), with the added attraction of additional water in seasonally dry habitats (dilute nectar or juicy summer fruits) or it may provide a source of protein and other nutrients needed for growth or reproduction (pollen in flowers, nutrient-rich fruits like those consumed by specialist tropical frugivores) (McKey, 1975; Howe & Smallwood, 1982). Deception or mimicry is sometimes involved. Immature flowers and fruits must be protected against predators and against premature visits by vectors. The essential parts of the flowers (ovaries) or fruits (seeds) must not be damaged by the animal vector. Ineffective or parasitic visitors, which consume the reward without acting as vectors, must be discouraged.

Despite these extensive parallels, there are very important differences, both qualitative and quantitative, between zoophilous pollen dispersal and endozoochorous seed dispersal (Wheelwright & Orians, 1982). Seeds are much larger than pollen grains, usually far too large to be dispersed by flying insects; indeed, the comparatively small size of most frugivorous birds places an ecologically important limit on the size and mass of bird-dispersed seeds. In contrast to the short seasonal life cycles of many flower-visiting insects, frugivorous vertebrates are normally long-lived, so cannot be restricted to seasonally available fruit as a food source; obligate, fully specialized frugivory is possible only in tropical habitats. Populations of frugivorous vertebrates tend to be unstable and subject to cycling, so may be scarce or absent when needed. Effective frugivores may fail to disperse to, or only rarely visit, areas that are otherwise suitable for a fruit-dispersed species. Seed dispersal by frugivores, mediated by regurgitation or excretion, is often patchy and uncertain, in contrast to the efficiency and high specificity of pollen dispersal by many pollinators (Wheelwright & Orians, 1982). The seeds themselves are a potentially rich source of food, uneasily balanced by the food reward of the fruit in which they are contained; many seed-eating animals will consume the seeds but leave the fruit, and partial or complete digestion of both seeds and fruit by frugivores is quite common (Janzen, 1971). Mast fruiting in response to pressure from seed predators may, in a species with edible fruit, also adversely affect frugivore populations; Fleming, Breitwisch & Whitesides (1987) state that it is unknown in species with edible fruit in Africa and the neotropics, although community-wide mast fruiting does occur in areas of Malaysia with unpredictable dry seasons and biennial fruiting is common elsewhere in the Orient and Australasia. Frugivorous seed dispersal thus has many intrinsic disadvantages, which tend to limit its occurrence to habitats and ecological niches in which its potential advantages are greatest.

The distribution of frugivory among animal groups

Frugivorous seed dispersers occur in four of the seven major groups of vertebrates; teleost fish, reptiles, birds and mammals (van der Pijl, 1982; Willson, 1983). The first two groups are only of very minor importance.

Frugivory and frugivorous seed dispersal by fish has been studied mainly in tropical habitats. In Amazonia, the fruits of a range of riverside trees are reported to be attractive to fish, and some are used as fish bait (van der Pijl, 1982). Gottsberger (1978) found the seeds of a total of 33 species of plant in the stomachs of 12 larger fish species from Amazonia; the seeds of 16 plant species were germinable. Goulding (1981) found fruit-eating as well as seed-eating in several Amazonian fish species, and reported that some of the fruit-eating fish might be important seed dispersers. The fleshy fruits of the Indonesian mangrove *Sonneratia* are eaten by fish, which may act as seed dispersers, although the chief dispersers (and pollinators) of *S. alba* are bats (van der Pijl, 1982). Frugivory by European fishes appears to have received little attention. Many freshwater aquatic plants are known to be dispersed very effectively by waterbirds, or within lakes and river systems by water (van der Pijl, 1982); fish dispersal seems likely to be at most incidental or of very minor importance in freshwater habitats in Europe.

Among the relatively small number of modern reptiles that are chiefly or partly herbivorous, several examples of frugivory have been reported. One of the best-known is the dispersal of the Galápagos tomato *Lycopersicon esculentum* var. *minor* by the tortoise *Testudo elephantopus*; seeds take 12–20 days to pass through the gut but show enhanced rates of germination (Rick & Bowman, 1961). Large tortoises (*Geochelone gigan-*

tea) also act as seed dispersers on Aldabra (Hnatiuk, 1978). The fleshy berries of *Podophyllum peltatum*, a common woodland herb in the eastern United States, are eaten and its seeds effectively dispersed by box turtles (*Terrapene carolina*) (Rust & Roth, 1981), which have good colour vision and are reported to eat a range of fruits (Klimstra & Newsome, 1960). Reptiles including the smaller dinosaurs are likely to have been of importance as frugivores during the Mesozoic period, when they may have dispersed fruit-bearing gymnosperms like *Ginkgo* (Janzen & Martin, 1982) and perhaps early angiosperms (Tiffney, 1984; Herrera, 1989; Wing & Tiffney, 1987).

Birds form the most important group of frugivores, especially in temperate regions. The syndrome of adaptations associated with bird-dispersed fruits is well-known. Birds have excellent sight, with colour discrimination extending into the near-UV in some cases (Burckhardt, 1982) but have a weak sense of smell; they can climb and fly, so can reach fruits inaccessible to ground-dwelling animals. They are relatively small in size so unable to ingest or carry very large seeds, and have no teeth so cannot easily penetrate or chew hard fruit coats, although they are able to grind seeds in their gizzards after ingestion. Fruits adapted to bird dispersal thus typically have (van der Pijl, 1982, with modifications):

1, an attractive edible part;
2, when immature, an outer, green and distasteful (often acid) protection against premature eating;
3, an inner protection of the seed against digestion (hard seed-coat or endocarp, bitter or toxic compounds in seed);
4, signalling colours when mature, e.g. orange, red, blue, purple, black or UV-reflecting white, often in contrasting combinations;
5, not necessarily any smell;
6, a permanent attachment to the plant;
7, a conspicuous position on the plant;
8, no closed, hard fruit coat;
9, relatively small size, usually less than 20 mm in diameter.

There are exceptions to most of these generalizations. Mimetic 'fruits' formed by hard seeds with fruit-like appearance and position obviously have no edible part; several certain or possible examples, e.g. the legume *Adenanthera pavonina*, are cited by van der Pijl (1982). Immature fruits are sometimes red, as in *Rubus fruticosus*, and may then form part of a bicoloured display contrasting with mature fruit (Willson & Thompson, 1982; Willson & Melampy, 1983; Greig-Smith, 1986; Janson, 1987). The seeds may be adapted for regurgitation rather than excretion, and then may lack protection against digestion. The seeds within fruits are often eaten by seed predators, and are sometimes consumed together with the fruits, evidently lacking effective chemical defence. Snow & Snow (1988) have suggested that mechanical and chemical protection of the seed are alternatives; plant families have tended to adopt either mechanical or chemical protection, but not both. This is not always so, as for example in *Prunus* where mechanically well-protected seeds may contain cyanogenic glucosides, but may often be the case. Specialized seed-eaters are, however, likely to tolerate seed toxins.

Frugivorous birds are most abundant and diverse in tropical habitats. Fleming, Breitwisch & Whitesides (1987) found that there were 405 frugivorous (defined as having a diet at least 50% of which is composed of fruits) bird species in the Neotropics, 149 in Africa and 143 in Southeast Asia. Tropical frugivorous groups like the cotingas (Cotingidae) include some extreme fruit specialists (Snow, 1982); bellbirds, for example, feed both themselves and their young wholly or predominantly on fruits. McKey (1975) has drawn attention to the ways in which the fruits eaten by specialist avian frugivores differ from those eaten by non-specialists; the former have firm flesh, rich in fats and proteins, and large seeds; the latter have succulent flesh rich in carbohydrate, with smaller seeds. Some large frugivorous birds of tropical forests, for example toucans and hornbills, can pluck and swallow fruits which are often 30 mm or more in diameter and up to a maximum of about 40 mm × 70 mm (Snow, 1981); these fruits are huge in comparison with the rather small size, often less than 10 mm in diameter, of most bird-consumed fruits. Other smaller frugivorous tropical birds have extremely wide gapes, which permit them to swallow fruits that are huge for the size of the bird (Snow & Snow, 1988). Fruit-producing plants are correspondingly abundant in tropical forest, with fruiting periods extending through the year, which enable them to support specialized frugivores (van der Pijl, 1982). The diversity of frugivorous birds does, however, appear to be limited in comparison with that of insectivorous birds, perhaps because fruits form conspicuous displays in a limited number of ways whereas insects are ecologically diverse and avoid predation in many different ways. Snow (1976) pointed out that among neotropical frugivorous birds there are 79 species of Cotingidae (cotingas) and 59 species of Pipridae (manakins), compared with 215 species of the purely insectivorous Furnariidae (ovenbirds) and 222 species of the insectivorous Formicariidae (antbirds).

Mammals are also important frugivores, especially in tropical habitats where specialized frugivory is widespread in bats and primates. Fleming, Breitwisch & Whitesides (1987) found that there were 96 frugivorous (see above) species of bat and 33 frugivorous arboreal species of primate in the neotropics; the corresponding numbers in Africa were 26 bat species and 32 arboreal primate species, and in Southeast Asia 66 bat species and 11 arboreal primate species. Mammals differ from birds in several ways, which affect their behaviour and effectiveness as frugivores. Mammals perceive fruits in a different manner from birds; unlike birds, most mammals are wholly or partly nocturnal and have an acute sense of smell, but are colour-blind. Primates, which have good colour vision in Old World taxa, weaker in New World species, and are often specialist or habitual frugivores, are an exception. Mammals also process fruits in different ways; they often have strong jaws and teeth, so can penetrate hard fruit coats; they may crush seeds while chewing fruits, but lack the grinding gizzards of birds. Ruminants and other large herbivorous mammals may have protracted digestive processes, and tend to produce comparatively large droppings in which large numbers of seeds may be aggregated. Fruits adapted for dispersal by colour-blind mammals tend to be whitish, yellowish, greenish or dull in colour, but often have a strong or characteristic smell. Tropical bats are the only flying group of mammalian frugivores; fruit-bats tend to be rather larger in body size than most frugivorous birds, so are able to handle comparatively large fruits. They typically squeeze juice from the fruit and regurgitate or otherwise discard the seeds; fruits adapted for dispersal by fruit-bats are a comparatively well-defined class (van der Pijl, 1982).

Non-flying mammalian frugivores may be wholly or partly arboreal; if terrestrial, they are limited to fruits that are shed from trees or bushes (for example apples, the fruits of *Malus* spp., and quinces, the fruits of *Cydonia* spp.) or to fruits that grow in accessible positions near to the ground. Terrestrial mammalian frugivores vary greatly in size, from small rodents or marsupials weighing only a few grams at one extreme to large herbivores weighing hundreds of kilograms at the other. Elephants are known to be active frugivores, seeking and eating a range of fruits, which may have very large seeds (e.g. *Dumoria heckeli* and *Mangifera*) (van der Pijl, 1982). Elephants are reported by Alexandre (1978) to be the chief dispersers of up to 30% of the forest trees in the Ivory Coast, seeds germinating rapidly from elephant dung. Elephants and other African herbivores disperse the seeds of many leguminous trees and shrubs, eating their leathery, nutritious pods and excreting the hard seeds. The Mediterranean carob tree, *Ceratonia siliqua*, is a European example of this type of dispersal adaptation. A disadvantage of seed dispersal by large herbivores is the extreme patchiness of the primary seed shadow. Lamprey, Halevy & Makacha (1974) describe a Tanzanian example in which an elephant dropping weighing 8 kg was found to contain 12 000 *Acacia tortilis* seeds. Large herbivores may, however, deposit seeds in areas where grazing, browsing, dunging and trampling have created favourable conditions for the establishment of seedlings. It has been suggested by Janzen & Martin (1982) that the widespread extinctions of large herbivorous mammals during the late Pleistocene may have resulted in the loss of the natural dispersal mechanisms of some tree species. They pointed out that in Central America, where the extinction of a megafauna including horses, gomphotheres and giant sloths coincided with the spread of the Paleo-Indians about 10 000 years ago (Martin, 1973), the large fleshy fruits of a number of native tree species (e.g. *Enterolobium cyclocarpum*, guanacaste) now commonly remain undisturbed below the trees. They suggested that these trees were dispersed by extinct herbivores. Janzen (1981) found that the seeds of *E. cyclocarpum* survived fairly well when they were fed to (reintroduced) range horses; most seeds were excreted within 14 days, but some not until much later (two horses excreted the last seeds after 30 and 33 days, and a third horse retained some for 60 days); overall viabilities after excretion ranged from 17 to 56%.

Primates, apart from man now almost entirely confined to tropical habitats, are common frugivores, but sometimes act as seed predators (Snodderly, 1979) or are rather ineffective or wasteful as seed dispersers (Ridley, 1930). Monkeys, for example, were found to be major consumers of the fruits of *Tetragastris panamensis* on Barro Colorado Island in Panama, accounting for 88% of the viable seeds removed from the plants (Howe, 1980) but excreted most seeds in aggregations, which germinated to give dense clusters of seedlings from which only one plant seemed likely to survive; in addition, they ate fruits of this species mainly in years of unusually high fruit production, so were not reliable dispersers. Primates may eat the fruits of a very wide range of plants; Lieberman *et al.* (1979) found germinable seeds of 59 species in the dung of free-living baboons in Ghana. Edible fruits form a significant and highly valued, though usually not essential, part of the food of nearly all human cultures, but the number of species that have achieved wide distribution through endozoochory in Man appears to be surprisingly small, although the tomato, *Lycopersicon esculentum*, is an obvious example, and *Opuntia* spp. (prickly pears) can also be dispersed in this way (Ridley, 1930).

The frequency and ecological distribution of endozoochory in the European flora

Fruit-bearing trees and shrubs are a familiar sight in Europe, especially in agricultural and suburban landscapes, but they form a surprisingly small proportion of the native flora as a whole. Estimates of the proportions of species in different evolutionary and ecological categories in the European flora are complicated by the inconsistent treatment of apomictic and autogamous complexes, e.g. *Hieracium, Rubus fruticosus* s.l. and *Taraxacum*. Some of these taxonomically difficult complexes have been subdivided into very large numbers of microspecies by specialists. If these microspecies are included in an overall enumeration and given the same weighting as normal sexual species, the proportions of groups in which apomixis is rare will be underestimated or distorted (Kay & Stevens, 1986). In the standard taxonomic account of the European flora, *Flora Europaea* (Tutin *et al.*, 1964–80) most apomictic complexes have been divided into groups of representative microspecies. These groups are variously designated as collective species, series, sections, etc. but generally correspond to sexual species in their genecological range and morphological distinctiveness. Taking these groups as species-equivalents, the best estimate of the number of native seed plants (angiosperms and gymnosperms) in the area covered by *Flora Europaea* is 10 409 (Table 1), in 146 families. Of these, only 390 species (3.8% overall; 3.8% of dicots and 2.8% of monocots) (Table 2) in 45 families have edible fruits adapted for endozoochory. Among gymnosperms, succulent 'fruits' are surprisingly frequent, occurring in 14 of the 34 native species (41.2%). Small fruits predominate, with 47.6% having mean diameters between 6 and 9 mm and a further 27.2% between 10 and 14 mm, but a significant number, 11.0%, have fairly large fruits with mean diameters in greater than 20 mm (Table 3). The commonest fruit colours are red (37.5%) or blue-black to black (41.5%), but a wide range of fruit colours occurs, including two blue-fruited species, *Lonicera caerulea* and *Viburnum tinus* (Table 4). The great majority of the fruit-bearing species in the European flora are plants of woodland, maquis (Mediterranean scrub) or dwarf-shrub moorland communities. Most are small trees, shrubs, climbers or woodland herbs. Succulent fruits are rare in plants of grass-dominated and ruderal communities, which form a large proportion of the flora of Europe.

In the British Isles, fruit-bearing trees and shrubs are particularly conspicuous in the hedges and woodland margins of the traditional agricultural countryside, but again form only a small proportion of the native flora, although rather greater than in Europe as a whole, probably because woodland and moorland species are proportionately more abundant in Britain. Of the 1380 native seed plant species or species-equivalents (Kay & Stevens, 1986) described by Clapham, Tutin & Warburg (1962), 87 (6.3% overall; 7.6% of dicots and 2.7% of monocots) have edible fruits adapted for endozoochory (Tables 1 and 2). Two of the three native gymnosperms produce succulent fruits. Most have comparatively

226   Q.O.N. KAY

Table 1. *Frequencies of seed plants with endozoochorous fruits among native species in four regional floras*

The numbers of families and species with endozoochorous fruits are shown in parentheses.

	Families	Species	% of species with endozoochorous fruits
British Isles[1]	112 (26)	1380 (87)	6.3
Europe[2]	146 (45)	10409 (390)	3.7
Canary Islands[3]	65 (22)	547 (54)	9.8
Sydney region[4]	141 (57)	1935 (237)	12.2

Source: [1]Clapham, Tutin & Warburg, 1962; [2]Tutin *et al.*, 1964–80; [3]Bramwell & Bramwell, 1974; [4]Beadle, Evans & Carolin, 1972.

Table 2. *Frequencies of endozoochory among different groups of seed plants in the native British and European floras, compared with the native flora of the Sydney region*

The numbers of families and species with endozoochorous fruits are shown in parentheses.

	Families	Species	% of species with endozoochorous fruits
British Isles[1]			
Gymnosperms	3 (2)	3 (2)	(66.7)
Dicotyledons	85 (20)	975 (74)	7.6
Monocotyledons	24 (4)	402 (11)	2.7
Europe[2]			
Gymnosperms	4 (3)	34 (14)	41.2
Dicotyledons	117 (38)	8640 (328)	3.8
Monocotyledons	25 (4)	1735 (48)	2.8
Sydney region[3]			
Gymnosperms	2 (1)	11 (2)	(18.2)
Dicotyledons	105 (50)	1331 (219)	16.5
Monocotyledons	34 (6)	593 (16)	2.7

Source: [1]Clapham, Tutin & Warburg, 1962; [2]Tutin *et al.*, 1964–80; [3]Beadle, Evans & Carolin, 1972.

Table 3. *Mean diameters of endozoochorous fruits in four regional floras*

The percentages of species in each size class are shown.

	Size class (mm)					
	0–5	6–9	10–14	15–19	20–29	30+
British Isles[1] (*n*=73)	8.22	50.68	31.51	6.85	1.37	1.37
Europe[2] (*n*=191)	8.38	47.64	27.23	5.76	8.38	2.62
Canary Islands[3] (*n*=22)	18.18	27.27	27.27	4.55	22.73	–
Sydney region[4] (*n*=79)	21.52	35.44	25.32	5.06	6.33	6.33

Source: [1]Clapham, Tutin & Warburg, 1962; [2]Tutin *et al.*, 1964–80; [3]Bramwell & Bramwell, 1974; [4]Beadle, Evans & Carolin, 1972.

small fruits, with mean diameters of 5–15 mm; only two species (*Malus sylvestris*, crab-apple, and *Pyrus communis*, wild pear) have fruits above 20 mm in diameter (Table 3). The predominant fruit colours are red (41 species, 47.3%) or blue-black to black (36 species, 41.4%); there are no truly blue-fruited species (although several have a bluish waxy bloom, e.g. *Prunus spinosa* and *Vaccinium myrtillus*), only one with white fruits (*Viscum album*), two with greenish fruits and four with brown or brownish-orange fruits (Table 4). Nearly all are either woodland plants – small trees, shrubs, climbers or woodland herbs – or are plants of dwarf-shrub moorland communities (e.g. *Vaccinium* spp. and *Empetrum nigrum*). No dominant tree in native forest in the British Isles is endo-zoochorous although several of the dominant or formerly dominant tree species have comparatively small seeds, e.g. *Tilia cordata*, *Ulmus glabra*, *Betula* spp. and *Alnus glutinosa*. In contrast to this, some of the berry-bearing dwarf shrubs of moorland communities are frequent community dominants, e.g. *Vaccinium myrtillus* and *Empetrum nigrum*.

The flora of the Canary Islands is partly derived from the Pliocene forest that covered much of Europe before the Pleistocene glaciations, and makes an interesting comparison with the present European flora. Of the 547 native and mainly endemic Canarian species described by Bramwell & Bramwell (1974), 54 (9.9%) have succulent fruits adapted for endozoochory (Table 1). Many of these are laurel-forest species; in the

Table 4. *Colours of endozoochorous fruits in four regional floras*

The percentages of species in each colour category are shown.

	White	Green	Yellow	Brown	Orange	Red	Purple	Blue	Blue-black	Black
						Colour of mature fruit				
British Isles[1] (n=87)	1.15	2.30	0.0	4.60	3.45	47.13	0.0	0.0	9.20	32.18
Europe[2] (n=323)	3.10	1.24	6.19	4.95	3.72	37.46	1.24	0.62	4.64	36.84
Canary Islands[3] (n=38)	6.58	0.0	3.95	9.21	13.16	23.68	2.63	2.63	5.26	32.89
Sydney region[4] (n=61)	11.48	6.56	1.64	0.0	11.48	27.87	6.56	8.20	3.28	22.95

Source: [1]Clapham, Tutin & Warburg, 1962; [2]Tutin *et al.*, 1964–80; [3]Bramwell & Bramwell, 1974; [4]Beadle, Evans & Carolin, 1972.

laurel forests, which are thought to resemble the Pliocene vegetation of southern Europe, the proportion of endozoochorous species is much higher, including most of the dominant trees. Although the sample size is small, fruit size is probably more evenly distributed with both very small and large fruits being more frequent than in Britain and Europe (Table 3: five of the 22 species for which dimensions are stated have mean diameters above 20 mm). Fruit colour also appears to be more diverse, with fewer red fruits and more white and purple or blue fruits than in Europe (Table 4). A larger sample size and more accurate comparative data are needed. Endemic pigeons, able to handle comparatively large fruit and likely to respond to purple and blue fruit displays, are important frugivores in the Canary Islands and may have spread laurel forest trees and shrubs to the outer islands, including Tenerife, by long-distance dispersal, thus being partly responsible for the survival of the Canarian flora in its island refuges but also biasing the forest flora towards bird-dispersed endozoochory. The native avifauna is small but includes the frugivorous *Sylvia* warblers *S. atricapilla* and *S. melanocephala*. In Britain *S. atricapilla* mainly consumes small fruits, less than 7 mm in diameter (Snow & Snow, 1988), as in Spain, although in Spain they will also consume comparatively large (though small-seeded) fruits like those of *Ficus carica* and *Arbutus unedo* (Jordano & Herrera, 1981).

The Sydney region of Australia, extending from the sclerophyllous heathland and evergreen forest remnants of the coast into the Blue Mountains, is a well-studied and ecologically fairly diverse unglaciated area with a warm-temperate climate comparable to the pre-Pleistocene climate of Europe, and a rich avifauna. Beadle, Evans & Carolin (1972) describe 1935 native seed plant species in the area. Of these, at least 237 (12.3%) have succulent fruits adapted for endozoochory (16.5% of dicots but only 2.7% of monocots). Bird dispersal predominates; frugivorous birds are more numerous and diverse than in Europe. Most of the fruits are small or very small, 21.5% with mean diameters of 5 mm or less and 60.9% with mean diameters of 6–14 mm, but a significant number (12.7%) have mean diameters in excess of 20 mm (Table 3). As in the Canaries, fruit colour is apparently more evenly distributed than in the British and European flora, with comparatively high proportions of white and purple or blue fruits (Table 4). The large-fruited component may include a few rainforest species adapted to fruit bats.

Comparisons between the British and overall European floras show that relatively small, red or black fruits predominate in both, with closely similar proportions of species in the four smaller size classes (Table 3); the main difference in size range is the greater frequency of medium to large fruits (mean diameter greater than 20 mm) in the European flora

(11.0% in Europe, compared with 2.7% in the British Isles). In colour, blue-black or black fruits have very similar frequencies (41.5% in Europe, 41.4% in Britain), although red fruits appear to be rather less numerous in Europe as a whole (37.5% in Europe, 47.1% in the British Isles). The chief differences in colour are the greater frequency of white and especially yellow fruits in Europe (there is only one white-fruited species and no predominantly yellow-fruited species in the British Isles, but they make up 9.3% of European endozoochorous fruits) and the occurrence of a few purple and blue-fruited species in Europe, again absent from the British Isles. The differences in colour may result from the relative impoverishment of the British flora rather than from the absence of some dispersal syndromes from Britain; in British gardens, the ripe fruits of introduced species with yellow, purple and blue fruits are often consumed eagerly by native frugivorous birds, especially Blackbirds (*Turdus merula*). Both in Britain and in Europe as a whole, most endozoochorous fruits are evidently adapted for primary dispersal by birds, and indeed for dispersal by a comparatively small range of birds (Snow & Snow, 1988); the small size of the fruits, their red, black or other contrasting colours, their retention on the fruiting plants in conspicuous displays and the seasonal phasing of their production are among their adaptations for bird dispersal. Adaptations for endozoochorous dispersal by mammals are most likely to be represented among the larger edible fruits, often hard-fleshed and with mean dimensions greater than 20 mm. These are more frequently greenish or brown, and are often shed from the parent plant before dispersal (which commonly fails under modern European conditions); most of the few European examples are in the Rosaceae–Pomoideae, including some *Malus* spp. (apples), some *Pyrus* spp. (pears) and *Mespilus germanica* (medlar), but *Ceratonia siliqua* (carob) in the Leguminosae also appears to be adapted for this type of dispersal.

The taxonomic distribution of endozoochorous fruits

Plants with fruits adapted for endozoochorous dispersal are very unevenly distributed among different families of seed plants in the European flora (Table 5). Although edible fruits occur in 390 species distributed among 45 families, the single family Rosaceae has 164 species (or species-equivalents; see above) with edible fruits, 42.1% of the modest (Table 1) European total. The Liliaceae, with 28 species with edible fruits, and the Caprifoliaceae with 22 lag far behind, although with six more families (Araliaceae, Rhamnaceae, Thymelaeaceae, Solanaceae, Ericaceae and Paeoniaceae) they account for a further 34.4%. The Rosaceae also contribute the greatest number of species to the endo-

Table 5. *Families with endozoochorous fruits in the native flora of Europe*

The number of native European species with endozoochorous fruits is shown for each family, with the numbers native in the British Isles in parentheses. Nomenclature follows Tutin *et al.* (1964–80) and the order of the families follows Stebbins (1974).

Gymnosperms			Myrtaceae	1
Cupressaceae	9	(1)	Cornaceae	4 (2)
Taxaceae	1	(1)	Elaeagnaceae	1 (1)
Ephedraceae	3		Santalaceae	3
			Loranthaceae	3 (1)
Dicotyledons			Rafflesiaceae	2
Lauraceae	2		Celastraceae	5 (1)
Berberidaceae	4	(1)	Aquifoliaceae	1 (1)
Ranunculaceae	2	(1)	Rhamnaceae	17 (2)
Myricaceae	1		Vitaceae	1
Caryophyllaceae	1		Anacardiaceae	7
Chenopodiaceae	1		Cneoraceae	1
Paeoniaceae	10		Zygophyllaceae	1
Hypericaceae	1	(1)	Araliaceae	1 (1)
Ulmaceae	4		Oleaceae	6 (1)
Moraceae	1		Solanaceae	13 (3)
Cucurbitaceae	2	(1)	Verbenaceae	1
Capparidaceae	2		Rubiaceae	5 (1)
Ericaceae	11	(8)	Caprifoliaceae	22 (6)
Empetraceae	2	(2)		
Myrsinaceae	1		**Monocotyledons**	
Rosaceae	164	(33)	Palmae	1
Grossulariaceae	9	(5)	Araceae	18 (2)
Leguminosae	1		Liliaceae	28 (8)
Thymelaeaceae	15	(2)	Dioscoreaceae	1 (1)

zoochorous component of the native British flora, in which the 33 species with edible fruits in the Rosaceae make up 37.9% of the British total of 87 species with edible fruits. Other families contribute much smaller numbers; the Liliaceae s.l. and the Ericaceae each have eight edible-fruited species in Britain, the Caprifoliaceae six, the Grossulariaceae five and no other family more than three. The predominance of the Rosaceae in Britain and Europe appears not to be due to the proliferation and narrower circumscription of species among apomictic (*Rubus* and *Sorbus*) or subsexual (*Rosa*) sections of edible-fruited genera in the family. The microspecies in these genera have been grouped into collective species-equivalents to make comparisons as realistic as possible (see above); non-apomictic genera among European Rosaceae also contribute large numbers of edible-fruited species, e.g. *Prunus* (12 species), *Pyrus* (11 species) and *Crataegus* (20 species). In the British flora, members of the Rosaceae are often conspicuously the most abundant fruiting species in hedgerows and scrub in the autumn; *Crataegus monogyna, Prunus spinosa, Rosa* spp. and *Rubus* spp. in the lowlands, with *Prunus padus* and *Sorbus aucuparia* in the uplands. Some edible-fruited species in other families are also abundant, e.g. *Sambucus nigra*, but except in dwarf-shrub moorland the numerical predominance of the Rosaceae is usually clear. In continental Europe, the Rosaceae are often similarly predominant outside the Mediterranean region, but within this region are less proportionately abundant among edible-fruited species.

In the Canary Islands there are comparatively few edible-fruited Rosaceae; Bramwell & Bramwell (1974) describe only six native species (four endemic *Bencomia* spp., *Rubus* cf. *ulmifolius* and *Prunus lusitanica*). For comparison, they describe ten Liliaceae and four Lauraceae among the total of 54 species with edible fruits. These are distributed among 22 families, several of which are now predominantly tropical (e.g. Santalaceae, Lauraceae, Sapotaceae and Myrsinaceae). A similar tropical element occurs among edible-fruited Mediterranean species, for example members of the Moraceae (*Ficus*), Santalaceae (*Osyris*), Rafflesiaceae (*Cytinus*), Laureaceae (*Laurus*), Cneoraceae (*Cneorum*), Anacardiaceae (*Pistachia*), and some Oleaceae (*Olea* and *Phillyrea*). European plants with edible fruit show much greater diversity in Continental Europe than in the British Isles at the family level (Table 5) largely because this Mediterranean element with tropical relationships is absent from the British flora.

In the Sydney region, species with edible fruits are distributed among a much greater variety of families than in Europe (Table 6). The Epacridaceae, with 38 edible-fruited species (16.0% of the total), forms the largest single component but does not approach the numerical predominance shown by the Rosaceae in Europe; it is perhaps more comparable to European Ericaceae, reflecting the importance of the sclerophyllous coastal heathlands in the Sydney flora. The prominent Southern Hemisphere family Proteaceae has 17 edible-fruited species in the region, and the specialist Loranthaceae (which are almost entirely dependent on bird dispersal) 15; the remaining species with edible fruits are distributed comparatively evenly among a relatively large number of families with predominantly tropical affinities. This pattern of distribution is probably typical for endozoochory in warm-temperate and tropical climates.

Table 6. *Families with endozoochorous fruits in the native flora of the Sydney region of Australia*

The number of native species with endozoochorous fruits is shown for each family. Nomenclature follows Beadle, Evans & Carolin (1972) and the order of the families follows Stebbins (1974).

Gymnosperms		Epacridaceae	38	Araliaceae	3
Podocarpaceae	2	Sapotaceae	1	Apocynaceae	1
		Ebenaceae	2	Oleaceae	5
Dicotyledons		Myrsinaceae	2	Solanaceae	10
Winteraceae	2	Cunoniaceae	7	Ehretiaceae	1
Eupatomiaceae	1	Pittosporaceae	2	Verbenaceae	3
Monimiaceae	6	Rosaceae	4	Myoporaceae	3
Lauraceae	8	Escalloniaceae	2	Gesneriaceae	1
Cassythaceae	4	Thymelaeaceae	1	Lobeliaceae	5
Piperaceae	3	Myrtaceae	7	Goodeniaceae	4
Menispermaceae	3	Proteaceae	17	Rubiaceae	6
Amaranthaceae	1	Santalaceae	2	Caprifoliaceae	2
Chenopodiaceae	3	Olacaceae	5		
Symplocaceae	2	Loranthaceae	15	**Monocotyledons**	
Elaeocarpaceae	5	Celastraceae	2	Flagellariaceae	1
Ulmaceae	2	Icacinaceae	2	Palmae	2
Moraceae	8	Euphorbiaceae	2	Araceae	4
Violaceae	1	Rhamnaceae	2	Lilliaceae	3
Flacourtiaceae	1	Vitaceae	5	Smilacaceae	4
Passifloraceae	2	Anacardiaceae	1	Philesiaceae	2
Capparidaceae	1	Rutaceae	3		

The ecology of endozoochory in Europe

In the modern European flora, birds are by far the most important agents involved in primary dispersal of endozoochorous fruits. The greatest contributions to our knowledge of the extent and ecological importance of avian frugivory in Europe have come from the work of Herrera & Jordano in the Iberian Peninsula (Herrera & Jordano, 1981; Herrera, 1981*a,b*, 1982*a,b*, 1984*a,b*; Jordano, 1982, 1985, 1987*b*; Herrera, 1987, 1988) and from the work of Barbara and David Snow in England (Snow & Snow, 1988). Neither of these areas is typical of the European temperate forest in which most non-Mediterranean European species with endozoochorous fruits were originally native. In the Iberian Peninsula, temperate forest is mainly montane and predominantly divided up into comparatively small relict areas, while in England only a few small vestiges of the original forest cover remain and forest species with endo-zoochorous fruits survive mainly in a man-made rural landscape of hedgerows, plantations and secondary scrub (Rackham, 1986). The Mediterranean forests and scrublands studied by Herrera & Jordano are also partly anthropogenic and much affected by human activities. Fortunately, avian frugivores are mobile and often adaptable to changing circumstances, but it should be remembered that the relationships between frugivores and fruit-bearing plants in Europe are often closely dependent on human activities and artificial modification of the landscape.

In the area of the British Isles studied by Snow & Snow (1988) only fourteen species of bird that also act as seed-dispersers regularly ate wild fruits: five thrushes, three mainly resident and two winter migrants; the mainly resident European Robin; four *Sylvia* warblers, chiefly summer migrants; the Starling, resident but much augmented by migrants from other parts of Europe during the winter; and three crows. Woodpigeons (*Columba palumbus*) also frequently acted as dispersers of the seeds of fleshy fruits although they have grinding gizzards and also act as seed predators. Four finches regularly acted as seed predators on fleshy fruits, and three tits acted either as pulp-predators or seed-predators (Table 7). At least 39 species of plant with fleshy fruits were native in their study area in lowland England. The frugivorous birds were often observed to feed mainly on different fruits in the field, different bird species thus often showing strong apparent preferences for one or more fruiting species and differing orders of preference. These apparent preferences were not necessarily confirmed in situations where choice was possible. Song Thrushes, for example, showed a strong apparent preference for the red fruits of *Viburnum opulus*, which were avoided by other thrushes, and when offered a choice did strongly prefer them to the red fruits of *Crataegus monogyna*. Fieldfares fed mainly on the fruits of *Rosa* spp. and *Crataegus monogyna*, and also strongly preferred them in choice situations. Redwings, however, fed mainly on the fruits of *Ilex aquifolium*, *Cornus sanguinea* and *Rhamnus catharticus*, but in choice situations preferred the fruits of *Crataegus monogyna*. The feeding spectrum and preferences shown by the thrushes appeared to depend on interactions between availability in the preferred habitat of the frugivore (*Crataegus monogyna*, for example, is particularly abundant in hedges in the open country favoured by Fieldfares), palatability, and competitive interactions between species (Mistle Thrushes, for example, are larger than the other thrushes and will defend fruiting *Ilex aquifolium* trees against the others, often successfully; Redwings form mixed flocks with the larger Fieldfares which may prevent them from eating *Crataegus monogyna* fruits).

Table 7. *Birds observed to eat fruits or act as seed predators on edible fruits in central southern England*[1]

R, resident; SM, summer migrant; WM, winter migrant.

Seed-dispersers	
Turdus merula (Blackbird)	R + WM
T. philomelos (Song Thrush)	R (+WM)
T. viscivorus (Mistle Thrush)	R (+WM)
T. pilaris (Fieldfare)	WM
T. iliacus (Redwing)	WM
Erithacus rubecula (European Robin)	R
Sylvia atricapilla (Blackcap)	SM (+WM)
S. borin (Garden Warbler)	SM
S. communis (Common Whitethroat)	SM
S. curruca (Lesser Whitethroat)	SM
Sturnus vulgaris (Starling)	R + WM
Corvus corone (Carrion Crow)	R
Pica pica (Magpie)	R
Garrulus glandarius (Jay)	R
Seed disperser and predator	
Columba palumbus (Woodpigeon)	R
Seed predators	
Pyrrhula pyrrhula (Bullfinch)	R
	R (+WM)
Carduelis chloris (Greenfinch)	
	R
Fringilla coelebs (Chaffinch)	
F. montifringilla (Brambling)	WM
Seed or pulp predators	
Parus major (Great Tit)	R
P. caeruleus (Blue Tit)	R
P. palustris (Marsh Tit)	R

Source: [1]Snow & Snow (1988).

In the area of lowland England studied by Snow & Snow (1988), the Vale of Aylesbury, fruiting trees and shrubs, with some fruiting woodland herbs, are most frequent in hedgerows and on the edges of surviving woods. Their fruits are generally well suited for dispersal by the comparatively small number of avian frugivores in the area, with which they could be regarded as having a common relationship. Fruit production begins in June, early to mid-summer (e.g. *Fragaria vesca, Prunus avium, Rubus idaeus, Ribes* spp.), peaks during September (e.g. *Crataegus monogyna, Prunus spinosa, Rubus fruticosus* agg., *Sambucus nigra, Sorbus aria*) but continues through the autumn and winter so that some fruits may still be available in spring (e.g. *Hedera helix, Ilex aquifolium, Viscum album*). This seasonal pattern of fruit production appears to correspond with the seasonal abundance and feeding specializations of potential frugivores. In mid- to late summer, resident populations of thrushes, Starlings, etc. and summer migrant populations of *Sylvia* warblers peak in England. Their protein requirements at this time are lower than during the breeding season and fruits provide an acceptable food. Migrants begin to pass through during September, and winter migrant populations of thrushes and Starlings arrive during the autumn in large numbers from other parts of Europe, staying in Britain until the spring and feeding extensively on fruits. Similar seasonal correlations of fruit availability with the presence of frugivores have been described in other areas of the world. In the eastern deciduous forests of North America, where the winters are very severe and all frugivorous birds are migratory, the seasonal production of bird-dispersed fruits and the latitudinal patterns of fruit abundance closely coincide with the southward passage of migrant frugivores (Stiles, 1980). In southern Spain, as Herrera (1982a, 1984a) and Jordano (1982, 1985) have shown, the fruiting seasons of many plants appear to be adapted to the seasonal presence of wintering populations of migrant frugivorous birds from northern Europe, peaking in late autumn and winter. Snow & Snow (1988) point out that in Europe, as Stiles (1980) has shown in eastern North America, fruiting seasons should thus be earliest in the north – perhaps peaking in July–August in northern Scandinavia – and increasingly late towards the south.

Fruit quality also shows seasonal correlations. Summer fruits, produced when water may be required in comparatively large amounts by frugivorous birds during hot, dry weather, tend to contain a high proportion of water, with low seed mass and a very watery pulp, both in England (Snow & Snow, 1988) and in Mediterranean habitats (Herrera, 1982a). Water content decreases and relative yield increases in fruits produced later in the year, and energy-providing lipid content peaks in winter. The fruits of *Hedera helix* (ivy), with 31.9% lipid content, and of *Olea europaea* (olive) with 41.9% lipid content (dry mass basis in both cases) (Herrera, 1987) are particularly striking examples of lipid-rich winter fruits.

The frugivore–fruit relationships studied in southern England by Snow & Snow (1988) were essentially those found in the temperate woodland

and woodland-edge communities of western Europe, and are likely to represent a single diffusely adapted coadapted complex (see below). The other non-Mediterranean European community in which plants with endozoochorously dispersed fruits are frequent is dwarf-shrub moorland, in which fruits are produced during the summer by several shrubs, which may be community dominants, e.g. *Vaccinium myrtillus*, *V. vitis-idaea* and *Empetrum nigrum*, as well as (in the British Isles) less abundant species like *Rubus chamaemorus* and *Cornus suecica*. These fruits are eaten and their seeds dispersed by grouse and ptarmigan (*Lagopus* spp.) and related herbivorous game-birds, which may also disperse the unadapted seeds of other dwarf shrubs in the community, consuming the seeds of the latter with the foliage (Ridley, 1930). This moorland frugivore–fruit relationship clearly represents a second coadapted complex distinct from that of temperate woodland in Europe.

Studies of the relationships between frugivores and fruit-producing plants have often emphasized the ecology of the frugivores and the benefits that they obtain from frugivory, perhaps partly because it is easier to quantify these than to study the costs of fruit production and the patterns and effectiveness of fruit and seed dispersal, and partly because many studies have been basically ornithological. The close ecological correlations of endozoochorous fruit production with particular plant habitats (woodland and scrubland, dwarf shrub communities) and strategies within those habitats (small trees and shrubs of understorey or clearings, hemiparasitic epiphytes, perennial woodland herbs) and its near-complete absence from most other plant habitats show that dispersal by endozoochorous fruits is likely to involve comparatively substantial costs which are counteracted by specific and substantial benefits in the habitats where it occurs. These benefits may include:

(a) dispersal of fruits or seeds to or within the specialized habitat of the fruit-producing plant, e.g. by a habitat-specific disperser;

(b) placement of fruits or seeds in a suitable germination site within that habitat, perhaps inaccessible by other means (e.g. epiphytic hemiparasitic Loranthaceae on tree branches, chasmophytic *Cotoneaster* and *Sorbus* spp. on cliffs);

(c) dispersal of fruits or seeds away from the seed source (escape) or over long distances (colonization of new or ephemeral habitats, e.g. forest clearings);

(d) dispersal into a regular pattern, reducing intraspecific competition (e.g. by defecation or regurgitation of seeds at constant time intervals along a foraging route);

(e) dispersal by species-constant frugivores separately from potential competitors;

(f) dispersal at a time of year or in a habitat when or where other means of dispersal, e.g. wind, are not available, or into habitats that they cannot reach (e.g. summer-fruiting deciduous woodland shrubs, scrubland lianes, evergreen forest trees);

(g) dispersal to sites that have been modified by the animal disperser in a way that increases chances of successful germination and establishment (e.g. sites grazed, trampled and/or fertilized by mammalian dispersers, sites fertilized by roosting birds);

(h) protection of the seeds from seed predators during and after dispersal (Herrera, 1982b);

(i) modification of characteristics of protective seed-coat, etc. to enhance germination after dispersal, e.g. *Lycopersicon esculentum* (Rick & Bowman, 1961).

There have been few studies in which these possible benefits have been tested or demonstrated, especially in temperate habitats, although several have shown specific benefits and adaptations in relationships between frugivores and mistletoes (e.g. Reid, 1989). Seedfall patterns in bird-dispersed species were studied by Hoppes (1988) who showed that in an Illinois woodland seeds were dispersed preferentially into clearings. Preferential dispersal of the seeds of *Prunus mahaleb* by its frugivorous avian dispersers (*Turdus merula* and *Sylvia atricapilla*) towards the sites that were apparently safest for the growth and survival of its saplings was observed by Herrera & Jordano (1981). Fallow deer (*Dama dama*) and red foxes (*Vulpes vulpes*) consume fallen fruits of *P. mahaleb*; Herrera & Jordano observed that the deer seemed to destroy all the seeds that they ingested, but intact seeds were seen in red fox excreta, suggesting that they may act as minor dispersal agents. Red foxes commonly eat blackberry (*Rubus fruticosus* agg.) fruits in southern Britain, leaving characteristic purple-stained scats containing large numbers of seeds.

The adaptations of endozoochorous plants may involve secondary ecological costs. Most winter-fruiting species, for example, have seed dormancy mechanisms delaying seed germination until at least a year after seed production (Snow & Snow, 1988), perhaps because dispersal may be delayed until it is too late for successful germination to take place in the first spring, but increasing the chance of fruit or seed loss to pests or seed predators. Mast fruiting (the production of large seed crops only in certain years, often simultaneously by different tree species, to reduce the effects of fruit pests and seed predation) might be thought to be reduced or absent in endozoochorous species because it would seriously affect frugivore populations. However, in a study of interactions between fruit production, avian seed dispersal and levels of fruit predation by insect pests in the wild olive, *Olea europaea* var. *sylvestris*, Jordano (1987b) found that fruit production varied greatly from year to year, almost completely failing in the first year of the study but producing a glut in the second year. In the first year, levels of fruit destruction by the insect pests were low (*ca.* 6%) and almost all (96%) of the few fruits that were produced were dispersed by avian frugivores, of which there were fourteen main species. In the second year of glut, the pest populations rapidly increased to track fruit production, and levels of fruit destruction by insect pests were much higher (up to 52% with a mean of 27%) but excess fruit production satiated the bird dispersers, who removed only 52% of the fruit, although a greater absolute number of seeds were dispersed because of the larger crop. Thus in this case neither obvious benefits nor apparent disadvantages were associated with mast fruiting.

Correlations between endozoochory, dioecy, pollination biology and defence by thorns

A number of authors have pointed out that dioecy is unusually frequent in gymnosperm and tropical angiosperm species with endozoochorous fruits, especially those with bird dispersal (Bawa, 1980; Givnish, 1980, 1982; Willson, 1983). Both Givnish and Bawa have suggested that separation of male and female functions may enable dioecious species to allocate a greater proportion of resources to the production of nutritious fruits and conspicuous fruit displays. In the British flora, there is a clear association between dioecy and endozoochory: 13 of the 87 native endozoochorous species are dioecious (14.9%), in comparison with an overall frequency of 4.4% of dioecious species in the native seed plant flora as a whole (Kay & Stevens, 1986). However, this may result from an association between endozoochory and growth-form, because dioecy, like endozoochory, is most frequent among small trees and shrubs. There is an extremely strong correlation between endozoochory and insect pollination; in the British flora, 82 (94.3%) of native endozoochorous species are insect-pollinated, a much greater proportion than in the flora as a whole. This may be associated both with the open and scattered population structure that is often produced by endozoochorous dispersal and with the sheltered woodland habitat and late flowering seasons of many endozoochorous species, for which wind pollination is inappropriate.

One of the most striking correlations with endozoochory in the European flora is its association with defence by thorns or prickles. Among the endozoochorous species in the native British flora, 30 (34.5%) are thorny or prickly. This obviously confers defence against grazing and browsing herbivores in many endozoochorous woody species (e.g. *Berberis vulgaris* and many Rosaceae); the alternative form of defence by toxic chemical constituents occurs in many herbaceous endozoochorous species (e.g. *Actaea spicata*, *Arum maculatum*, *Atropa belladonna*, *Convallaria majalis*). However, it is also possible that thorniness or prickliness may be selectively advantageous in endozoochorous trees and shrubs because it protects the fruits against inappropriate frugivores (e.g. terrestrial mammals in bird-dispersed species) and also because it protects avian frugivores from predators while they are foraging. Birds also tend to roost in thorny trees or shrubs, regurgitating or defecating seeds to the ground below while they roost; this may or may not be advantageous.

Ecological genetics and evolution of endozoochory in Europe

The relationships between frugivores and plants producing endo-zoochorous fruits have sometimes been assumed to be closely coevolved, partly by analogy with coevolved pollination syndromes (Snow, 1971; McKey, 1975; Thompson, 1982). However, Wheelwright & Orians (1982) and others have pointed out that the interactions between fru-givores and fruits differ from those between pollinators and flowers in a number of important ways; in particular, plants will normally gain more advantage from a diffuse, generalist relationship with frugivores than from a highly specific one. In this type of relationship, groups of organisms with similar adaptations, rather than single species, are involved on both sides (Herrera, 1982a; Wheelwright, 1985; Jordano, 1987a; Snow & Snow, 1988). Fruiting species in a diffuse coevolved relationship will thus tend to have similar signal characteristics (e.g. colour, shape and position of fruit) and handling characteristics (e.g. seed size, texture and digestive properties of the pulp); frugivores will have similar visual acuity, searching and perhaps fruit-handling behaviour, and similar or compatible metabolic needs and beak or jaw structure.

Within a group of frugivores or fruiting plants involved in a diffusely coevolved relationship, facilitative or competitive interactions may lead to the sharing of resources in space or time. The seasonal phasing of fruit maturation, so that the same group of frugivores (one or more species, usually several) is supported by a temporal succession of different plant species with more or less similar fruits, is one likely outcome of this type of interaction. Another possible outcome is the support of frugivores by groups (guilds) of simultaneously fruiting species in which some or all of the species do not produce enough fruits to support or attract the fru-givores on their own. Examples of both outcomes are described by Thompson & Willson (1979) from temperate forest in the eastern United States, where there is a summer guild of asynchronously ripening fruiting species adapted to the sparse population of resident frugivores and an autumn guild of simultaneously fruiting species adapted to the autumn peak of migrant frugivores. Both situations have close analogies in pollinator–flower relationships (Rathcke, 1983; Kay, 1987) and occur in the same temperate woodland environments. In tropical environments, temporal displacement of flowering and fruiting seasons as a result of competition for the services of dispersers has been described by Wheelwright (1985). Within the British Isles, it could be argued that today only two (temperate woodland and dwarf-shrub moorland) diffuse coevolved relationships between frugivores and fruiting plants exist, both involving birds as primary dispersers and probably separated into temporal guilds as in eastern North America, with one more specific relationship (*Viscum album*, mistletoe, and *Turdus viscivorus*, Mistle Thrush). In continental Europe and in Mediterranean habitats the same two diffuse relationships may exist, perhaps with other diffuse relationships involving different groups of birds (e.g. pigeons), as well as some more specific interactions including the mistletoe relationship (perhaps in more than one version) and also vestiges of some different and less evenly coevolved relationships involving mammals (see below). Further studies are likely to reveal a variety of more precise relationships within the overall context, perhaps in particular localities or under particular circumstances. An example of an unexpected frugivorous relationship is the discovery by Greenberg (1981) that two North American migrant *Dendroica* warblers, insectivorous in their summer range, appear to be specific frugivorous seed dispersers of tropical *Miconia* and *Lindackeria* trees in their winter range in Central America.

One of the most compelling arguments for the lack of selection for reciprocal specificity in frugivore–fruit relationships is the extreme conservatism of some fruits, especially fleshy fruits (Herrera, 1986). The fruits of the yew, *Taxus baccata*, for example, appear to be closely similar to those of *Palaeotaxus*, a taxad of the Upper Triassic, which existed nearly 200 million years ago, long before the appearance of modern birds, but yew fruits were the preferred choice of several of the thrush frugivores observed by Snow & Snow (1988). Fruits of modern Lauraceae are very similar to those of Eocene members of the family, but are eaten and dispersed by many modern frugivores. In both examples it is clearly the frugivores (now usually birds in both cases) that have adapted to the fruit, which has remained essentially unchanged. Structural adaptations to frugivory in habitually frugivorous European birds have been described by Herrera (1984*b*) and Jordano (1987*b*). The apparent conservatism of fruit adaptations may, however, partly result from the existence of a relatively small number of possible models or basic types of endozoochorous fruits, especially in diffuse coevolved relationships involving mobile and adaptable birds as the frugivores. In many genera (e.g. *Lonicera*) and even within some species (e.g. *Hedera helix*, which has yellow-fruited and black-fruited sub-species) fruits with different colours, shapes or other properties exist in closely related taxa, showing that selection can produce sharp changes in the characters of edible fruits.

Within a diffuse coevolved relationship, there is good evidence that natural selection has acted to produce intraspecific differentiation in fruiting characters in some European species. In *Smilax aspera*, a dioecious climber common in southern Spain, Herrera (1981a) found that berries may contain one, two or three seeds, the one-seeded berries having the highest pulp:seed ratio and thus being most rewarding for frugivorous birds. He found that the mean number of seeds per berry in local populations had a genetic basis and was inversely correlated with the availability of frugivorous dispersers, being lowest (and thus most rewarding) where frugivores were sparse or where there was competition for their services. Another example of intraspecific genetic variation in fruiting characteristics was reported by Valdeyron & Lloyd (1979) in the wild fig, *Ficus carica*. In European populations, two crops of figs, one of which (the brebas crop) normally fails, are produced annually by female figs; in more southerly populations, three crops are produced. Figs are important food resources for tropical frugivores, and have been suggested to be a key central resource (Terborgh, 1986) although this suggestion has not been supported by field observations in tropical Africa (Gautier-Hion & Michaloud, 1989).

Within the limits of the common fruit characters involved in a single diffuse relationship, there is considerable scope for variation and adaptation in seed mass, number and shape, and in characters that may affect the preferences of particular frugivore species, e.g. taste and nutritional quality. Frugivore behaviour may also be affected by the occurrence of cathartic secondary products in the fruit, as perhaps in *Rhamnus catharticus*, by the occurrence of seeds that are toxic (e.g. *Euonymus europaeus*) or unpalatably hairy (e.g. *Rosa canina*) so tend to be voided rapidly, or by adhesive constituents as in *Viscum album*.

Ecotypic or clinal genetic variation, often in populations which are quite closely adjacent to one another, has long been known to be common in plant species; more recently, studies of allozyme variation have shown that many plant species have sharply differentiated mosaic patterns of genetic variation even within populations (Levin & Kerster, 1974; Loveless & Hamrick, 1984; Lack & Kay, 1987). These patterns of variation result partly from high coefficients of selection and partly from the extremely small neighbourhood size (often less than 50 individuals within an area of a few square metres) consequent upon the limited pollen flow and low seed dispersal that are found in many species. With small neighbourhood sizes gene-flow may be very slow, and genetic differentiation produced by local selection or stochastic variation will persist. In endozoochorous species with wide seed dispersal, neighbourhood size should be far larger than the norm for plants, and high rates of gene-flow within and between populations are likely to eliminate most stochastic effects and reduce the overall extent of genetic differentiation. In species that tend to be dispersed unidirectionally by migrating birds, for example the woodland species of eastern North America studied by Stiles (1980) and perhaps similar woodland species in continental Europe, gene-flow by seed dispersal may oppose natural selection for adaptation to local conditions. Comparative studies using allozyme comparisons and other molecular techniques are needed.

The history of endozoochory in Europe

During the late Cretaceous and Eocene periods the vegetation of Europe was essentially tropical in character. The climate apparently cooled gradually through the Tertiary, until by the late Pliocene warm-temperate forests still including many species of tropical affinities grew in suitable areas of western, central and southern Europe. The geography and ecology of many parts of Europe were modified, sometimes drastically, by events associated with the Alpine orogeny during the Tertiary, when ranges of mountains were formed around the Mediterranean basin. In the Mediterranean basin itself, hot arid conditions must have existed when the Mediterranean sea dried to isolated hypersaline remnants during the Pliocene, followed by drastic changes when the Straits of Gibraltar were breached and the Mediterranean re-filled. Climates of the distinctly Mediterranean type seem to have appeared about three million years ago, after this re-filling, associated with a vegetation including tropical-margin elements (Raven, 1973). The frugivore–fruit associations that would have developed during the Tertiary in Europe are thus likely to have been ones of a generally tropical type, involving a range of specialized or habitual frugivores including both birds and mammals. Associations involving migratory birds were probably of much less importance, with their summer and autumn phases initially limited to the far north and east of the region, although they would have increased in extent and significance during the Pliocene. These associations often involve members of the Rosaceae, a largely non-tropical family. During the late

Pliocene the frugivore–fruit associations of tropical types would have become increasingly limited to the south and associated with Mediterranean vegetation.

European climates and ecosystems entered a period of particularly drastic change during the Pleistocene, with the onset of the Pleistocene glaciations (Frenzel, 1968). The glaciations were preceded by a number of cyclical cool periods of similar length during which adaptable plants and animals would have been at an advantage, with migratory animals and plants with good adaptations for long-distance dispersal being best suited to the unstable conditions. Selection during the cool periods may have increased the chances of survival for species and coadapted associations during the even more severe and fluctuating conditions of the four or more periods of full glaciation. The geography of Europe and North Africa is not well suited to north–south movement of communities during periods of climatic change, and many species must have become extinct or have been reduced to scattered remnants in Europe during the Pleistocene, including much of the tropical element (Leopold, 1967). Some of this element has survived in modern Mediterranean vegetation, and in a less modified form in a few outlying refuges such as the Canary Islands (see above).

Pleistocene plant and animal communities in Europe have ranged from warm-temperate forest, Mediterranean vegetation or steppe through temperate mixed or deciduous forest to boreal or montane coniferous forest and alpine fell-field or arctic tundra. They have been diverse and unstable in both time and space as a result of repeated climatic and geographic (sea-level) changes. These conditions must have favoured the development of diffuse coadapted relationships between frugivores and fruiting plants, especially those involving migrant birds, but perhaps in the past also those involving mobile or migratory herbivorous mammals. Specific associations between single species would be unlikely to survive unless, like some mistletoe–frugivore relationships, they were pre-adapted for mobility.

During the late Pleistocene a new factor emerged as Man began to modify the environment, both deliberately by the use of fire, clearance and eventually cultivation, and unconsciously by the reduction or extinction of ecologically important animals, or by the introduction of alien biota. Human or pre-human populations in Europe in the early Pleistocene were small, with little impact on the environment, although they are of interest as potential frugivores who may have been of some significance as minor dispersers of a few species, for example apples and similar fruits (see below). But as hunting techniques improved during the Devensian period it apparently became possible for Palaeolithic hunters to reduce populations of large herbivores such as horses and forest elephants drastically, and even to extinction (Martin, 1967, 1973). Thus Man may have been responsible for the late Palaeolithic extinctions of large herbivores and other megafauna in Europe from about 40 000 years before present. Some of these large herbivores may have been important frugivores and seed dispersers. Janzen & Martin (1982) have suggested that extinct gomphotheres, lost during megafaunal extinctions about 10 000 years ago, may have been the dispersers of *Enterolobium cyclocarpum* and other Central American species, which have large fruits apparently adapted for endozoochorous seed dispersal but consumed by no surviving native animals.

Can any corresponding element of endozoochorous plants specifically adapted for dispersal by extinct or drastically reduced megafauna be recognized in the European flora? The most likely group of European fruits apparently adapted for dispersal by comparatively large mammals are the larger fruits produced by members of the Rosaceae–Pomoideae (apples, pears, some large sorbs and medlars). These have several features associated with mammalian endozoochory: dull colour but often a distinctive scent, hard flesh or a hard skin, and dehiscence from the tree when mature so that they lie on the ground and can be eaten by terrestrial mammals. Horses eat apples eagerly and will seek them out, and are known to pass large numbers of uninjured seeds of a variety of species in their dung (Ridley, 1930). In Chile, introduced apples were reported to have become extensively naturalized in the south of the country as a result of dispersal by range cattle, their seeds germinating from cattle dung (Ridley, 1930); wild cattle were among the native megafauna of European deciduous forests. The extent to which surviving populations of large wild herbivores (mainly deer) and omnivores (wild pigs) could act as frugivorous seed dispersers in Europe is uncertain, but it seems likely that they are ineffective. Herrera & Jordano (1981) found that fallow deer (*Dama dama*) consumed the fruits of *Prunus mahaleb* but destroyed the comparatively well-protected seeds; Ridley (1930) reported no example of endozoochorous seed dispersal by temperate deer, and cited experiments by Kerner, which showed that pigs destroy the great majority of seeds that they consume. Domestic or feral cattle and horses may however still be able to act as seed dispersers for apples and similar fruits in many temperate areas of Europe, in addition to unconscious or accidental dispersal by Man. The carob (*Ceratonia siliqua*, Leguminosae–Caesalpinioideae) may have had a similar large-herbivore endozoochorous dispersal system and history in Mediterranean areas.

References

Alexandre, D.Y. (1978). Le rôle disséminateur des éléphants en forêt de Taï, Côte-d'Ivoire. *La Terre et La Vie* **32**, 47–72.

Bawa, K.S. (1980). Evolution of dioecy in flowering plants. *Annual Review of Ecology and Systematics* **11**, 15–39.

Beadle, N.C.W., Evans, O.D. & Carolin, R.C. (1972). *Flora of the Sydney Region*. Sydney: A.H. & A.W. Reed.

Bramwell, D. & Bramwell, Z. (1974). *Wild Flowers of the Canary Islands*. London: Stanley Thornes.

Burckhardt, D. (1982). Birds, berries and UV: a note on some consequences of UV vision in birds. *Naturwissenschaften* **69**, 153–7.

Clapham, A.R., Tutin, T.G. & Warburg, E.F. (1962). *Flora of the British Isles*. Cambridge University Press.

Faegri, K. & van der Pijl, L. (1979). *The Principles of Pollination Ecology*. Oxford: Pergamon Press.

Fenner, M. (1985). *Seed Ecology*. London: Chapman and Hall.

Fleming, T.H., Breitwisch, R. & Whitesides, G.H. (1987). Patterns of tropical vertebrate frugivore diversity. *Annual Review of Ecology and Systematics* **18**, 91–109.

Frenzel, B. (1968). The Pleistocene vegetation of northern Europe. *Science* **161**, 637–49.

Gautier-Hion, A. & Michaloud, G. (1989). Are figs always keystone resources for tropical frugivorous vertebrates? A test in Gabon. *Ecology* **70**, 1826–33.

Givnish, T.H. (1980). Ecological constraints on the evolution of breeding systems in seed plants: dioecy and dispersal in gymnosperms. *Evolution* **34**, 959–72.

Givnish, T.H. (1982). Outcrossing versus ecological constraints in the evolution of dioecy. *American Naturalist* **119**, 849–65.

Gottsberger, G. (1978). Seed dispersal by fish in the inundated regions of Humaitá, Amazonia. *Biotropica* **10**, 170–83.

Goulding, M. (1981). *The Fishes and the Forest*. Berkeley: University of California Press.

Greig-Smith, P.W. (1986). Bicolored fruit displays and frugivorous birds: the importance of fruit quality to dispersers and seed predators. *American Naturalist* **127**, 246–51.

Greenberg, R. (1981). Frugivory in some migrant tropical forest wood warblers. *Biotropica* **13**, 215–22.

Harper, J.L., Lovell, P.H. & Moore, K.G. (1970). The shapes and sizes of seeds. *Annual Review of Ecology and Systematics* **1**, 327–56.

Herrera, C.M. (1981a). Fruit variation and competition for dispersers in natural populations of *Smilax aspera*. *Oikos* **36**, 51–8.

Herrera, C.M. (1981b). Are tropical fruits more rewarding to dispersers than temperate ones? *American Naturalist* **118**, 896–907.

Herrera, C.M. (1982a). Seasonal variation in the quality of fruits and diffuse coevolution between plants and avian dispersers. *Ecology* **63**, 773–85.

Herrera, C.M. (1982b). Defense of ripe fruit from pests: its significance in relation to plant-disperser interactions. *American Naturalist* **120**, 218–41.

Herrera, C.M. (1984a). A study of avian frugivores, bird-dispersed plants, and their interactions in Mediterranean scrublands. *Ecological Monographs* **54**, 1–23.

Herrera, C.M. (1984b). Adaptation to frugivory of Mediterranean avian seed dispersers. *Ecology* **65**, 609–17.

Herrera, C.M. (1986). Vertebrate-dispersed plants: why they don't behave the way they should. In *Frugivores and Seed Dispersal* (ed. A. Estrada & T.H. Fleming), pp. 5–18. Dordrecht: W. Junk.

Herrera, C.M. (1987). Vertebrate-dispersed plants of the Iberian peninsula: a study of fruit characteristics. *Ecological Monographs* **57**, 305–31.

Herrera, C.M. (1988). The fruiting ecology of *Osyris quadripartita*: individual variation and evolutionary potential. *Ecology* **69**, 233–49.

Herrera, C.M. (1989). Seed dispersal by animals: a role in angiosperm diversification? *American Naturalist* **133**, 309–22.

Herrera, C.M. & Jordano, P. (1981). *Prunus mahaleb* and birds: the high-efficiency seed dispersal system of a temperate fruiting tree. *Ecological Monographs* **51**, 203–18.

Hnatiuk, S.H. (1978). Plant dispersal by the Aldabran giant tortoise *Geochelone gigantea* (Schweigger). *Oecologia* **36**, 345–50.

Hoppes, W.G. (1988). Seedfall patterns of several species of bird-dispersed plants in an Illinois woodland. *Ecology* **69**, 320–9.

Howe, H.F. (1980). Monkey dispersal and waste of a neotropical fruit. *Ecology* **61**, 944–59.

Howe, H.F. & Smallwood, J. (1982). The ecology of seed dispersal. *Annual Review of Ecology and Systematics* **13**, 201–28.

Janson, C.H. (1987). Bird consumption of bicolored fruit displays. *American Naturalist* **130**, 788–92.

Janzen, D.H. (1971). Seed predation by animals. *Annual Review of Ecology and Systematics* **2**, 465–92.

Janzen, D.H. (1981). *Enterolobium cyclocarpum* seed passage rate and survival in horses, Costa Rican Pleistocene seed dispersal agents. *Ecology* **62**, 593–601.

Janzen, D.H. & Martin, P.S. (1982). Neotropical anachronisms: the fruits the Gomphotheres ate. *Science* **215**, 19–27.

Jordano, P. (1982). Migrant birds are the main seed dispersers of blackberries in southern Spain. *Oikos* **38**, 183–93.

Jordano, P. (1985). El ciclo anual de los paseriformes frugivoros en el matorral mediterráneo del sur de España: importancia de su invernada y variaciones interanuales. *Ardeola* **32**, 69–94.

Jordano, P. (1987a). Patterns of mutualistic interactions in pollination and seed dispersal: connectance, dependence asymmetries, and coevolution. *American Naturalist* **129**, 657–77.

Jordano, P. (1987b). Avian fruit removal: effects of fruit variation, crop size and insect damage. *Ecology* **68**, 1711–23.

Jordano, P. & Herrera, C.M. (1981). The frugivorous diet of Blackcap populations *Sylvia atricapilla* wintering in southern Spain. *Ibis* **123**, 502–7.

Kay, Q.O.N. (1987). The comparative ecology of flowering. *New Phytologist* **106**, 265–81.

Kay, Q.O.N. & Stevens, D.P. (1986). The frequency, distribution and reproductive biology of dioecious species in the native flora of Britain and Ireland. *Botanical Journal of the Linnean Society* **92**, 39–64.

Klimstra, W.D. & Newsome, F. (1960). Some observations on the food coactions of the common box turtle, *Terrapene carolina*. *Ecology* **41**, 639–47.

Lack, A.J. & Kay, Q.O.N. (1987). Genetic structure, gene-flow and reproductive ecology in sand-dune populations of *Polygala vulgaris*. *Journal of Ecology* **75**, 259–76.

Lamprey, H.F., Halevy, G. & Makacha, S. (1974). Interactions between *Acacia*, bruchid seed beetles and large herbivores. *East African Wildlife Journal* **12**, 81–5.

Leopold, E.B. (1967). Late-Cenozoic patterns of plant extinction. In *Pleistocene Extinctions – the Search for a Cause* (ed. P.S. Martin & H.E. Wright), pp. 203–46. New Haven: Yale University Press.

Levin, D.A. & Kerster, H.W. (1974). Gene flow in seed plants. *Evolutionary Biology* **7**, 139–220.

Loveless, M.D. & Hamrick, J.L. (1984). Ecological determinants of genetic structure in plant populations. *Annual Review of Ecology and Systematics* **15**, 65–95.

Lieberman, D., Hall, J.B., Swaine, M.D. & Lieberman, M. (1979). Seed dispersal by baboons in the Shai Hills, Ghana. *Ecology* **60**, 65–75.

McKey, D. (1975). The ecology of coevolved seed dispersal systems. In *Coevolution of Animals and Plants* (ed. L.E. Gilbert & P.H. Raven), pp. 159–91. Austin: University of Texas Press.

Martin, P.S. (1967). Prehistoric overkill. In *Pleistocene Extinctions – the Search for a Cause* (ed. P.S. Martin & H.E. Wright), pp. 75–120. New Haven: Yale University Press.

Martin, P.S. (1973). The discovery of America. *Science* **179**, 969–74.

Matlack, G.R. (1989). Secondary dispersal of seed across snow in *Betula lenta*, a gap-colonizing tree species. *Journal of Ecology* **77**, 853–69.

Primack, R.B. (1987). Relationships among flowers, fruits and seeds. *Annual Review of Ecology and Systematics* **18**, 409–30.

Rackham, O. (1986). *The History of the Countryside*. London: Dent.

Rathcke, B. (1983). Competition and facilitation among plants for pollination. In *Pollination Biology* (ed. L. Real), pp. 305–29. London: Academic Press.

Rathcke, B. & Lacey, E.P. (1985). Phenological patterns of terrestrial plants. *Annual Review of Ecology and Systematics* **16**, 179–214.

Raven, P.H. (1973). Plant biogeography. In *Mediterranean Type Ecosystems* (ed. F. di Castri & H.A. Mooney), pp. 211–12. Berlin: Springer-Verlag.

Reid, N. (1989). Dispersal of mistletoe by honeyeaters and flowerpeckers: components of seed dispersal quality. *Ecology* **70**, 137–45.

Rick, C.M. & Bowman, R.I. (1961). Galápagos tomatoes and tortoises. *Evolution* **15**, 407–17.

Ridley, H.N. (1930). *The Dispersal of Plants throughout the World.* Ashford: Reeve.

Rust, R.W. & Roth, R.R. (1981). Seed production and seedling establishment in the mayapple, *Podophyllum peltatum*. *American Midland Naturalist* **105**, 51–60.

Snodderly, D.M. (1979). Visual discriminations encountered in food foraging by a neotropical primate: implications for the evolution of color vision. In *The Behavioral Significance of Color* (ed. E.H. Burtt), pp. 237–79. New York: Garland STPM Press.

Snow, D.W. (1971). Evolutionary aspects of fruit-eating by birds. *Ibis* **113**, 194–202.

Snow, D.W. (1976). *The Web of Adaptation: Bird Studies in the American Tropics.* London: Collins.

Snow, D.W. (1981). Tropical frugivorous birds and their food plants: a world survey. *Biotropica* **13**, 1–14.

Snow, D.W. (1982). *The Cotingas.* London: British Museum (Natural History).

Snow, B. & Snow, D. (1988). *Birds and Berries.* Calton: T. & A.D. Poyser.

Stebbins, G.L. (1974). *Flowering Plants: Evolution Above the Species Level.* London: Edward Arnold.

Stiles, E.W. (1980). Patterns of fruit presentation and seed dispersal in bird-disseminated woody plants in the Eastern deciduous forest. *American Naturalist* **116**, 670–88.

Terborgh, J. (1986). Keystone plant resources in the tropical forest. In *Conservation Biology: the Science of Scarcity and Diversity* (ed. M.E. Soule), pp. 330–44. Sunderland, Massachusetts: Sinauer Associates.

Thompson, J.N. (1982). *Interaction and Coevolution.* New York: Wiley.

Thompson, J.N. & Willson, M.F. (1979). Evolution of temperate fruit/bird interactions: phenological strategies. *Evolution* **33**, 973–82.

Tiffney, B.H. (1984). Seed size, dispersal syndromes and the rise of the angiosperms: evidence and hypothesis. *Annals of the Missouri Botanical Garden* **71**, 551–76.

Tutin, T.G., *et al.* (1964–80). *Flora Europaea*, vols. 1–5. Cambridge and London: Cambridge University Press.

Valdeyron, G. & Lloyd, D. (1979). Sex differences and flowering phenology in the common fig, *Ficus carica* L. *Evolution* **33**, 673–85.

van der Pijl, L. (1982). *Principles of Dispersal in Higher Plants*. Berlin: Springer-Verlag.

Wheelwright, N.T. (1985). Competition for dispersers and the timing of flowering and fruiting in a guild of tropical bird-dispersed trees. *Oikos* **44**, 465–77.

Wheelwright, N.T. & Orians, G.H. (1982). Seed dispersal by animals: contrasts with pollen dispersal, problems of terminology, and constraints on coevolution. *American Naturalist* **119**, 402–13.

Willson, M.F. (1983). *Plant Reproductive Ecology*. New York: John Wiley.

Willson, M.F. & Melampy, M.N. (1983). The effect of bicolored fruit displays on fruit removal by avian frugivores. *Oikos* **41**, 27–31.

Willson, M.F. & Thompson, J.N. (1982). Phenology and ecology of color in bird-dispersed fruits, or why some fruits are red when they are 'green'. *Canadian Journal of Botany* **60**, 701–13.

Wing, S.L. & Tiffney, B.H. (1987). Interactions of angiosperms and herbivorous tetrapods through time. In *The Origins of Angiosperms and their Biological Consequences* (ed. E.M. Friis, W.G. Chaloner & P.R. Crane), pp. 203–24. Cambridge University Press.

Index